T5-BAH-122

REAL-TIME MULTICOMPUTER SOFTWARE SYSTEMS

ELLIS HORWOOD SERIES IN COMPUTERS AND THEIR APPLICATIONS

Series Editor: IAN CHIVERS, Senior Analyst, The Computer Centre, King's College, London, and formerly Senior Programmer and Analyst, Imperial College of Science and Technology, University of London

Abramsky, S. & Hankin, C.J.	**ABSTRACT INTERPRETATION OF DECLARATIVE LANGUAGES**
Alexander, H.	**FORMALLY-BASED TOOLS AND TECHNIQUES FOR HUMAN–COMPUTER DIALOGUES**
Anderson, J.	**MODEL-BASED COMPUTER VISION: A Synthesised Approach**
Atherton, R.	**STRUCTURED PROGRAMMING WITH COMAL**
Baeza-Yates, R.A.	**TEXT SEARCHING ALGORITHMS**
Bailey, R.	**FUNCTIONAL PROGRAMMING WITH HOPE**
Barrett, R., Ramsay, A. & Sloman, A.	**POP-11**
Beardon, C., Lumsden, D. & Holmes, G.	**NATURAL LANGUAGE AND COMPUTATIONAL LINGUISTICS: An Introduction**
Berztiss, A.	**PROGRAMMING WITH GENERATORS**
Bharath, R.	**COMPUTERS AND GRAPH THEORY: Representing Knowledge for Processing by Computers**
Bishop, P.	**FIFTH GENERATION COMPUTERS**
Brierley, B. & Kemble, I.	**COMPUTERS AS A TOOL IN LANGUAGE TEACHING**
Britton, C.	**THE DATABASE PROBLEM: A Practitioner's Guide**
Bullinger, H.-J. & Gunzenhauser, H.	**SOFTWARE ERGONOMICS**
Burns, A.	**NEW INFORMATION TECHNOLOGY**
Carberry, J.C.	**COBOL**
de Carlini, U. & Villano, U.	**TRANSPUTERS AND PARALLEL ARCHITECTURES**
Chivers, I.D. & Sleighthome, J.	**INTERACTIVE FORTRAN 77: A Hands on Approach 2nd Edition**
Clark, M.W.	**PC-PORTABLE FORTRAN**
Clark, M.W.	**TEX**
Cockshott, W.P.	**A COMPILER WRITER'S TOOLBOX: How to Implement Interactive Compilers for PCs with Turbo Pascal**
Cockshott, W.P.	**PS-ALGOL IMPLEMENTATIONS: Applications in Persistent Object-Oriented Programming**
Colomb, R.M.	**IMPLEMENTING PERSISTENT PROLOG: Large, Dynamic, Shared Procedures in Prolog**
Cooper, M.	**VISUAL OCCLUSION AND THE INTERPRETATION OF AMBIGUOUS PICTURES**
Cope, T	**COMPUTING USING BASIC**
Curth, M.A. & Edelmann, H.	**APL**
Dahlstrand, I.	**SOFTWARE PORTABILITY AND STANDARDS**
Dah Ming Chiu, & Sudama, R.	**NETWORK MONITORING EXPLAINED: Design and Application**
Dandamudi, S.P.	**HIERARCHICAL HYPERCUBE MULTICOMPUTER INTERCONNECTION NETWORKS**
Dongarra, J., Duff, I., Gaffney, P., & McKee, S.	**VECTOR AND PARALLEL COMPUTING**
Drop, R.	**WORKING WITH dBASE LANGUAGES**
Dunne, P.E.	**COMPUTABILITY THEORY: Concepts and Applications**
Eastlake, J.J.	**A STRUCTURED APPROACH TO COMPUTER STRATEGY**
Eisenbach, S.	**FUNCTIONAL PROGRAMMING**
Ellis, D.	**MEDICAL COMPUTING AND APPLICATIONS**
Ennals, J.R.	**ARTIFICIAL INTELLIGENCE**
Ennals, J.R., *et al.*	**INFORMATION TECHNOLOGY AND EDUCATION**
Fillipic, B.	**PROLOG USER'S HANDBOOK**
Ford, N.	**COMPUTER PROGRAMMING LANGUAGES**
Ford, N.J., Ford, J.M., Holman, D.F. & Woodroffe, M.R.	**COMPUTERS AND COMPUTER APPLICATIONS: An Introduction for the 1990s**
Ford, N. & Ford, J.	**INTRODUCING FORMAL METHODS: A Less Mathematical Approach**
Gray, P.M.D.	**LOGIC, ALGEBRA AND DATABASES**
Grill, E.	**RELATIONAL DATABASES**
Grune, D. & Jacobs, C.J.H.	**PARSING TECHNIQUES: A Practical Guide**
Guariso, G. & Werthner, H.	**ENVIRONMENTAL DECISION SUPPORT SYSTEMS**
Harland, D.M.	**CONCURRENCY AND PROGRAMMING LANGUAGES**
Harland, D.M.	**POLYMORPHIC PROGRAMMING LANGUAGES**
Harland, D.M.	**REKURSIV**
Henshall, J. & Shaw, S.	**OSI EXPLAINED, 2nd Edition**
Hepburn, P.H.	**FURTHER PROGRAMMING IN PROLOG**
Hepburn, P.H.	**PROGRAMMING IN MICRO-PROLOG MADE SIMPLE**
Hill, I.D. & Meek, B.L.	**PROGRAMMING LANGUAGE STANDARDISATION**
Hirschheim, R., Smithson, S. & Whitehouse, D.	**MICROCOMPUTERS AND THE HUMANITIES: Survey and Recommendations**
Hutchins, W.J.	**MACHINE TRANSLATION**
Hutchison, D.	**FUNDAMENTALS OF COMPUTER LOGIC**
Hutchison, D. & Silvester, P.	**COMPUTER LOGIC**
Johnstone, A.	**LATEX CONCISELY**
Kirkwood, J.	**HIGH PERFORMANCE RELATIONAL DATABASE DESIGN**

Series continued at back of book

I/N.

REAL-TIME MULTICOMPUTER SOFTWARE SYSTEMS

RICHARD MARLON STEIN
Santa Clara, California, USA

ELLIS HORWOOD
NEW YORK LONDON TORONTO SYDNEY TOKYO SINGAPORE

First published in 1992 by
ELLIS HORWOOD LIMITED
Market Cross House, Cooper Street,
Chichester, West Sussex, PO19 1EB, England

A division of
Simon & Schuster International Group
A Paramount Communications Company

Printed and bound in Great Britain
by Bookcraft, Midsomer Norton

British Library Cataloguing in Publication Data

A Catalogue Record for this book is available from the British Library

ISBN 0–13–770777–0

Library of Congress Cataloging-in-Publication Data

Available from the publishers

Table of Contents

Dedication

To the loving memory of my fraternal and paternal grandparents: Edwin Morris and Beatrice Stalk, and David Nathaniel and Francis Hilda Stein.

To those who know what is not known.

Foreword

This text is a comfortable introduction to a complex and fascinating subject. The author's style is smooth and readable. His emphasis on software safety and the ethical use of multicomputer systems is well founded, and cannot be understated. This text affords a cognizant examination -- a snapshot in time really -- of an advancing technology from an active practitioner of multicomputer systems engineering. The technical descriptions and discussions are pertinent and colloquial. Anyone who is interested in becoming a practicing multicomputer software engineer with a bent for fast or real-time systems should peruse this text for a responsible and conscientious perspective of a continuously evolving scalable technology.

David L. Fielding
President, North American Transputer Users Group
Cornell University
Cornell, NY USA
January, 1992

Preface

As a vehicle for simulation and investigation, message-passing parallel computers -- multicomputer systems -- represent the most cost-effective problem solving tools yet invented. The scalable software built for these machines can provide insight and vision, but it can also imbue a nation with a strategic capability. Scalable multicomputer simulations can at once inform and enlighten, or destroy and pervert any purpose or conceivable idea created by mankind. Software applications are harder to construct in this realm of computation, and designing them requires unique skills and tools. The requisite engineering discipline must be acquired, practiced, and refined before effective and responsible use of a multicomputer platform can take place.

This text posits approaches to the solution of multicomputer software engineering problems. Among the questions explored by this text are: For what purpose have multicomputer systems been invented, and how are they applicable to the scientific and engineering problems which confront our society? What activities are necessary to prepare a proposal for a project based on a multicomputer system? What methods are available to estimate software engineering costs and schedule for a massively parallel simulation? How is software safety analysis used to prevent annoying or life-threatening mishaps from arising during simulation execution? How are real-time simulations engineered? What practices and skills are necessary to design and build a multicomputer simulation? How is a load balance realized? What methods are known for synchronizing a real-time multicomputer simulation? Answers to these questions are explored in the chapters of this book.

While the technical issues surrounding multicomputer software systems engineering are very important, they pale in comparison to the idea of a responsible multicomputer software engineering discipline. This state of mind is embraced and practiced by professional software engineers, and is expressed as a sincere concern for the societal implications of a finished edifice. Recognizing the likelihood of success or disaster hinges on the sensitive intuition of engineering judgement, a valuable characteristic of responsible practitioners. In contrast, the pure intellectual satisfaction of the engineering process is always visible, easier to appreciate and reward.

The end result of almost all engineering activities is a finished product which someone will use. The product may find a place in a home or a space shuttle. In either case, the consumer accepts that the product is safe, and will not harm the operator, or damage external equipment when used. The multicomputer software engineering discipline espoused in this text attempts to heighten the engineer's awareness of the principal steps involved while engineering a real-time multicomputer simulation. A real-time computer system requires disciplined software engineering skills to correctly design and build. Not only must the software correctly function in an algorithmic sense, but the results produced by the software must be temporally correct. A rare blend of skills are required to construct predictably correct real-time simulation software. Real-time multicomputer simulations are especially demanding in this respect.

With few exceptions, a real-time multicomputer simulation represents the technical pinnacle of software engineering accomplishment. The temporal coordination of tens, hundreds, or thousands of separate processing entities can be accomplished and harnessed for the benefit of mankind. Like so many soldiers on parade in perfect cadence, or the soothing instrumental notes and melodies of an orchestral arrangement, their actions and operations must be precisely controlled, regulated, and directed for an effective result to develop. Alternatively, the same platform can be driven with criminal intent, producing chaos and catastrophe for an overtly evil or amoral purpose. Multicomputer systems are not toys; they are seriously powerful instruments that can at once possess the destructive potential of a strategic weapon, or the humanitarian capacity of penicillin. The multicomputer software engineer must commit to a moral obligation and a code of ethics that restricts their practice to the safety and betterment of society.

Technology is changing at an exponentially scalable rate. The complexity produced from this accelerated evolutionary process is far beyond any individual's or any organization's capacity to recognize, control, or accurately contemplate. The societal and global ramifications resulting from this sustained and compounded introduction are profound. If technology obeyed Darwinian natural selection and survival of the fittest, the overpopulation of so many digital electronic devices would have died off by now, just like overpopulation of a species eventually exhausts a habitat's food supply, and equilibrium is restored through death by starvation. But technology is exempt from Darwinian law.

For the promulgation of software engineering safety awareness, and to promote recognition of the implicit strategic nature of multicomputer systems, I have attempted to combine and communicate my collectively acquired industrial, professional, and independent experience with both real-time simulation and multicomputer systems to you, the reader. Several months previous, I relished the thought of the technical challenge to describe a rapidly emerging field of research and investigation. But during the course of my background investigation, information assemblage, and revisions to the text, I developed a disquieting notion about this arena. This book is a distillation of my experience using multicomputer systems (chiefly Inmos transputer[1] boards which I purchased way back in 1985), and a recognition of the

[1] The transputer is a very practical device and affords an inexpensive avenue to start multicomputer investigations. My sincerest effort to avoid discussion of specific hardware products is represented here; I do not perceive this enthusiasm as an endorsement.

important and increasingly visible role they serve.

A democratic society has many rules and structures, such as the laws established by the United States Bill of Rights, to protect the citizenry and prevent societal breakdown. Technology does not obey the Constitution, but the ideas and freedom of expression conveyed through technology are principally sponsored by it. As a vehicle for expression, multicomputer technology wields immense power, and can be exploited for any purpose. With no legal or physical controls to regulate the advance of an unchecked specie, what will be the result? I do not know this answer, but it is very easy to assume to the worst. I hope that this text will serve as a baseline for the emerging practitioners of real-time multicomputer software systems engineering to consider the consequences of their work before undertaking design with impunity. I importune you to do so with caution and care.[2]

This text contains two principal sections. Part 1, Concepts and Practices, presents background material on multicomputer systems, project planning and preparation, software metrics, an introduction to real-time computer systems, and software safety. This collection of 5 chapters provides a foundation for software engineering practice and discipline. Part 2, Multicomputer Methods, supplies practical design information for scalable software engineering activities. Chapters that discuss software design, load balancing, and synchronization for multicomputer simulations are supplied. Each chapter is largely self-contained and decoupled from the others. They may be read in any order, depending on the individual's expertise and strength. Some background in software engineering is necessary and assumed. This book is not intended for the novice or freshman. However, it is targeted at the practicing professional software engineer with an eye toward real-world situations and experience.

Chapter 1 discusses multicomputer systems as a potential mechanism for yielding solutions to complex scientific, engineering, and bio-medical problems identified by the United States Federal High Performance Computing Program. Other nations, such as Japan, or the European Economic Community, have enacted similar programs to ensure a measure of technologically-derived economic security in the future. A brief discussion is given as to why multicomputers can address complex simulation issues, and are ideally suited to deliver cost-effective scalable solutions.

Chapter 2 discusses project planning and preparation. If one perceives the need to create a scalable multicomputer simulation solution, how is the project justified, a plan prepared and supported? The chapter is based on the presentation and discussion of a sample proposal written to the format solicited by the United States' Defense Advanced Research Projects Agency (DARPA).

Chapter 3 discusses software metrics. The COnstructive COst MOdel (COCOMO) developed by Barry W. Boehm is used as a platform to illustrate how a software project is estimated to quantify cost and schedule. Software engineering is more of an art than a science. Many factors affect the software lifecycle, the stages of planning, development, and maintenance that describe the useful lifetime of a software system from creation through retirement. Organizing an accurate estimate

---

[2] The author is painfully aware of this super-hypocritical argument -- a dilemma which scientists and engineers often confront.

of the engineering costs is far harder for a software product than, say, building an automobile or a toaster oven.

Two varieties of software are known for multicomputer simulations: data parallel and control parallel. On one hand, data parallel simulations are simply replicated instantiations of one (typically) simple process or calculation. This is true for many numerically intensive computations, like weather forecasting, finite element structure problems, or image processing; the same computation is conducted for each data element. In contrast, a control parallel simulation relies on a highly articulated process structure where many unique processes are combined to synthesize a simulated representation of a natural or man-made phenomenon. For each control parallel process, the argument is made to justify a separate engineering and cost estimate for lifecycle purposes. Control parallel simulations require multiple applications of software metric estimation to accurately cost.

Chapter 4 presents a capsulized introduction to real-time system simulation. Special attention is given to the predictability aspects of real-time systems. A real-time system must produce algorithmically or algebraically correct results which are also temporally correct. A real-time simulation is often a critical component of a larger system, such as an airplane, satellite, or an automobile. If the real-time system is not predictable under all circumstances, serious consequences can result in the event of system failure. The chapter outlines essential information on executive control structure, engineering tool requirements which are often found in real-time environments, and other common properties of this engineering discipline. The discussion is intended to impress an understanding and appreciation of real-time systems for their complexity, and give insight into some internal aspects of their implementation.

Software safety is the topic of Chapter 5. When good software goes bad, terrible things can happen which should not ever occur. Four examples of software failure are given. They have been reprinted from the pages of widely published newspapers, magazines, and journals. The strong and inseparable computer dependency seen in Western culture weaves at once a life-threatening and life-sustaining vein through our lives. Building software for use by others carries an implicit trust and ubiquity that is shattered each time the telephone does not work, or the ATM will not dispense money.

This chapter provides an overview of current software safety techniques and practice. Safety analysis techniques like Petri nets and software fault trees are introduced to show what kinds of methods are available to detect software faults before a failure occurs. While software safety has a terrific impact from the user's perspective, the engineering and design aspects are far more difficult to manage in a cost-effective way. Over 60% of all errors introduced into software arise from poor requirements definition or ambiguity. A discussion of formal specification methods is presented which argues for their adoption. If the requirement is poorly understood, it is likely that the final software implementation may also generate spurious functionality which can lead to disaster.

In Part 2, beginning with Chapter 6, multicomputer software design practice is discussed with an eye toward important design and implementation issues. To build multicomputer software, a different state of mind is necessary, one which is alien to sequential software engineering practices. A method of software design is presented which is derived from a process structure graph analysis representation of simulation. This is a familiar concept to anyone who has studied Hoare's Communicating Sequential Processes (CSP), or experienced the Occam model of programming on the Inmos transputer. The text explains how to construct this logically concurrent representation, simulate it within a comfortable environment to verify determinism, and then transform it to a physically concurrent multicomputer system. An example problem is provided which illustrates the concurrent thought process required to successfully carry out these operations.

A multicomputer runs most efficiently when all computation elements are equally loaded with data. The notion of load balance is introduced and examined in Chapter 7 along with a discussion of several techniques for realizing one. A partial taxonomy is created of the known and widely practiced load balancing methods. One technique is posited for real-time multicomputer simulations, while the majority are efficacious for purely static or quasi-dynamic problems.

The issue of temporal regulation and synchronization of the multicomputer simulation is the focus of Chapter 8. The distributed nature of multicomputer architecture poses unique problems for the temporal coordination of ten, one hundred, or several thousand computation elements, since each has it own clock source. A discussion of a synchronization algorithm is given that is applicable to homogenous multicomputer platforms.

Chapter 9 supplies a brief discussion of emerging trends and advanced technology likely to replace the existing generation of multicomputers. The parallel random access machine (PRAM) may well become the first general purpose parallel processing architecture, where the obstacles of load balance, topology, and data decomposition disappear from the multicomputer simulation scene.

Acknowledgments

I give my sincerest thanks to the staff at Ellis Horwood for their patience, prodding, and encouragement during the preparation of this manuscript. I am also grateful to Mr. Philip Presser, a long-time friend, mentor, and associate. I have been very fortunate to enjoy, experience, and absorb his thoughtful and oracle-like wisdom, patience, discipline, and insightful recognition of engineering methodology. Finally, and most importantly, I acknowledge the love and support from my family who have cared, nurtured, and raised their son, brother, nephew, and cousin with endearment and kindness.

Part I

Concepts and Practices

1

Why Multicomputers?

A technological revolution is a self-perpetuating system driven by evolutionary and innovative forces. Central to this theme is the computer. The most recent and advanced species is the message-passing concurrent computer: the multicomputer. Multicomputers embody the latest manifestation of computer systems engineering techniques and very large scale integrated (VLSI) circuit manufacturing processes into a scalable edifice, and this feature affords a substantial advantage in cost and performance. Multicomputers can be used to design and build ever more complicated and powerful systems, and discern solutions to problems and phenomena that their predecessors could not. Their existence substantiates the revolution and rejuvenates the evolutionary technological process. Multicomputers are vehicles that possess an earth-shaking potential for discovery.

This chapter lists and discusses applications that are suitable targets for multicomputer technology. The Grand Challenges of Computing, a set of 20 problems which serve as the principle motivation for the United States High Performance Computing Program [FHP89], are funded to investigate these avenues of research considered pivotal to the continued economic success of the United States.[3] These Challenges form the backdrop for our presentation. A discussion of multicomputer cost, sizing, memory, and performance issues is presented in this context.

1.1 Definitions

A *multicomputer* is defined as a computation system composed of two principle components: a homogeneous ensemble of processors with memory, and an interconnection network used to exchange messages between the processor ensemble. The most generic form of a multicomputer architecture schematic is shown in Figure 1.1. Each processor and memory combination is called a *node*. Each node is equipped with a separate timing source that regulates instructions and processor control.

General-purpose microprocessors typically serve as the central processor units in a node. It is possible and often desirable to substitute more exotic and less functional components in place of microprocessors, such as digital signal processor chips in the case of High Definition Television (HDTV) simulation. But throughout

---

[3] The High Performance Computing Initiative may be viewed as a calculated gamble to restore economic prosperity through the organization of an information-based economy.

this text, we confine the discussion about multicomputers to those based on general-purpose microprocessors.[4]

The interconnection network serves as a conduit of information transport. Messages pass through the network; they are transparently routed between processors on a non-interference[5] basis. The network may be composed of an Ethernet, Fiber Data Distributed Interconnect (FDDI), or specialized message-passing elements built from application-specific integrated circuits (ASICs). Throughout this text, special purpose routing chips are assumed to comprise the interconnection network architecture instead of more traditional network structures.

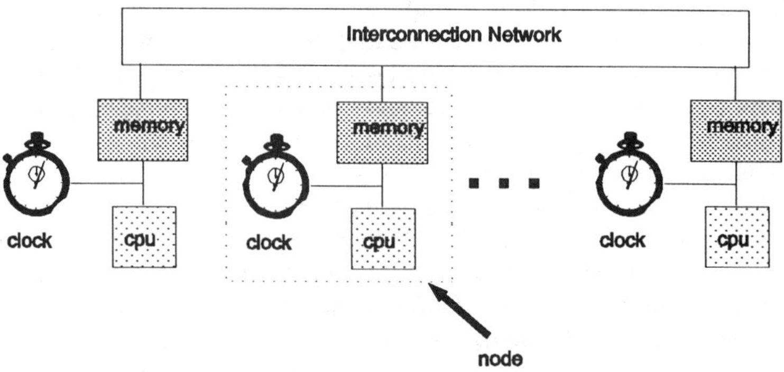

Figure 1.1 Generic multicomputer architecture schematic.

The message-passing feature of multicomputers implements an architecture which is largely contentionless. A contentionless computer system has no resource conflicts, and this feature permits the construction of an architecture which is *scalable*. A scalable architecture is constructed via multiple instantiations of identical nodes and message-passing support. Tens of thousands of nodes can be instanced and packaged

---

4 The MIMD-class (multiple instruction, multiple data) multicomputer is almost exclusively discussed in this text. The author has no experience with SIMD-class systems; they are not discussed.

5 Packet collisions are unavoidable. The author regards collisions and message-passing conveyance times as functions of hardware implementations that, in most cases, portend trivial consequences for a simulation.

together into a multicomputer platform.

A scalable architecture establishes the feasibility of implementing a scalable simulation and/or algorithm on the multicomputer. A scalable simulation is capable of continuous time-wise solution improvement in direct proportion to the number of nodes in the multicomputer. This is called *speed-up*. The speed-up for a simulation can be estimated from Amdahl's law ([Amd67] and [Fox88a]), which describes the portion of a sequential program which may be decomposed into many parallel parts that are executed on separate processors. Speed-up cannot always be achieved due to the underlying algorithmic dependence and computational domain structure (see Chapter 6) of the simulation. Alternatively, applying parallelism to the solution of a larger problem is called *scale-up*. Scale-up parallelism is often easier to extract from large problems, where a preponderance of data is divided among N processors, and the solution on each processor is conducted independently, save for the occasional communication between them. A solution is scaled-up to attack large problems on many processors, an approach that is often accomplished with the single-program multiple data (SPMD) programming style discussed in Chapter 6. Many applications discussed in this chapter are suitable for scale-up parallelism because the datasets found in their computational domains are huge.

A scalable simulation is achievable only when contention is architecturally absent from the computer platform. Contention confines the computer platform to an asymptotic performance metric, rather than a continuous functional rise with the addition of more processors to the computer system. Contention in shared-memory architectures arises from the competition between processors and their attempt to access a common resource: the memory bus connecting the processors to a global memory. The shared-memory multiprocessor architecture is contention-bound, whereas multicomputers are not.

If one compares the hypothetical execution speed of an ideal[6] simulation constructed to run on a shared-memory multiprocessor, and a multicomputer constructed with the same processor technology, a break-even point in the solution speedup will be seen. At this point, the multicomputer and shared-memory multiprocessor are computationally congruent. But the addition of more processors in the multicomputer case will produce an even faster solution, while the shared-memory system reaches an asymptotic level. This behavior is illustrated in Figure 1.2.

The rate of change in the multicomputer simulation speedup as a function of the number of processors applied to the problem is a constant value K. The shared-memory system also exhibits a linear growth for a small number of processors. Addition of processors beyond the break-even point in the shared-memory case produces a saturation in bus bandwidth, and the simulation speedup will reach a constant value. This characteristic is a universal artifact of shared-memory systems.

The line in Figure 1.2 that indicates the expected multicomputer performance improvement as a function of the number of nodes has a slope of:

---

[6] An ideal simulation means that the algorithm and data are easily decomposed and organized to permit a scalable multicomputer simulation.

$$\frac{dS}{dN} = K \qquad 1.1$$

But the shared-resource system shows signs of a bottleneck; it is most evident after the break-even point. The maximum speedup is limited by the availability of a finite resource, denoted as M, the maximum obtainable memory bandwidth in the presence of competition between processors. The slope of the shared-resource system is:

$$\frac{dS}{dN} = K(M - S) \qquad 1.2$$

and the solution to this differential equation is:

$$S(N) = M(1 - e^{-KN}) \qquad 1.3$$

This solution says that the expected speedup for a shared-resource system is approximately linear for a small number of processors, and progressively asymptotic for larger collections of processors. This heuristic derivation indicates why one almost never sees more than eight processors (four is typical) simultaneously operating on a problem in a shared-memory platform. Contention is a giant-killer.

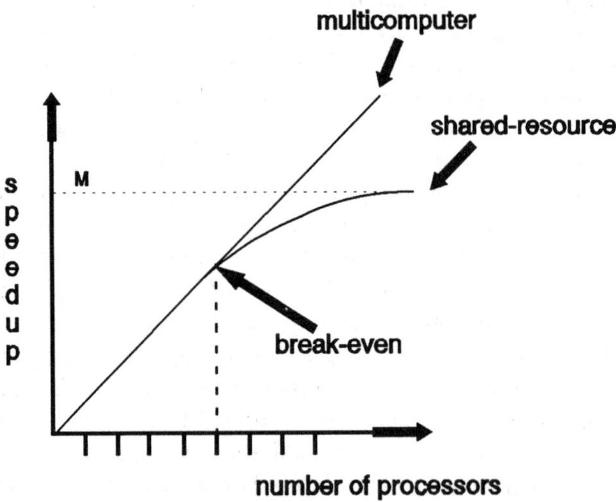

Figure 1.2 Heuristic derivation of asymptotic performance limit arising from contention in the shared-resource computer system. (Byte Magazine, June 1991 © McGraw-Hill, Inc., New York. All rights reserved).

Multicomputers free simulations from contention-bound performance increases, but distinct and peculiar issues for the software engineering arise which have no readily available automated solution or aid. It is more difficult to engineer a multicomputer simulation. But the roadblocks in multicomputer software engineering must be hurdled to clear the path for the resolution of crucial and important problems in many fields of science, medicine, and engineering.

1.2 Multicomputer Applications

Concurrent software, software targeted for multicomputer systems, is best constructed by participants who are well-versed and well-trained in the notion of *concurrent thought*. Concurrent thinkers possess the specialized knowledge, skill, and state of mind[7] to participate in the most challenging phases of a multicomputer software effort: specification, design, code, integration, and test. Calling upon a novice software engineer to build a multicomputer program invites disappointment, and is an ill-advised choice. Without the training and background, precious time and resource can be wasted while the novice discovers deadlock, builds a poor process structure, creates an awkward data decomposition, or struggles with a serious load imbalance. Concurrent software engineers possess the essential skills needed to realize an effective multicomputer simulation.

A multicomputer is not necessary to perform word processing, play video games, perform inventory control, or execute any other common and ordinary application. Sequential applications like these are constructed without much concern for process structure, load balancing, and other salient aspects of multicomputer simulation. Multicomputer applications are best suited for problem domains that defy expeditious solution and approximation via existing sequential systems. Multicomputer systems, through their scalable architecture and software structure, offer a realistic alternative to push the envelope of a problem -- to obtain a solution -- viewed as intractable for a sequential system into one that is reachable.

The issues of process structure, deadlock, load imbalance, and data decomposition are unique to multicomputer software systems. These issues are the barriers -- soft barriers -- that forestall the rapid rise of multicomputer technology into a ubiquitous agent of change. For example, if the software engineers who build today's computer-aided engineering programs were all trained and capable of building a multicomputer implementation of their products, a huge amount of money and time could be saved by automobile manufacturers during vehicle design and prototyping. Since a multicomputer system is generally 10 to 100 times more cost-effective[8] than an equivalent shared-memory platform, and the software which executes on the multicomputer is entirely scalable, then 10 to 100 times as many designs and prototypes

---

[7] The state of mind can resolve and visualize the additional aspects of software engineering unique to multicomputer simulation implementations.

[8] Based on a $/MIPS criterion with equal memory configuration.

can be examined for an equivalent investment.[9]

Many industries could be equally affected by multicomputer technology were it not for the pressing shortage of concurrent software engineers. Acquiring concurrent software engineering skills and multicomputer experience is unfortunately more of a privilege than a right, but only temporarily so. Universities in the United States, Japan, and Europe are slowly developing curricula for undergraduates that increasingly rely on parallel computation systems and algorithm design. But the general student body (undergraduates) are restricted from using parallel machines, and the multicomputer cycles are usually kept for research purposes. A select few commercial industries, such as oil companies, apply multicomputer technology for specialized purpose.

The defense industry and military installations are the principal users of multicomputer systems in the United States. The multicomputer is overtaking and replacing the shared-memory multiprocessor supercomputer -- the phallic symbols of computing. In the case of nuclear weapon implosion calculations, constructing a more precise shock-wave (which can increase the bomb yield by 1% or 2%) is a national pastime. These fluid calculations consume machine cycles like candy.[10] Only the scalable speed provided by multicomputers economically satisfies the demands of this problem class, and similar ones with equally voracious habits.

Molecular dynamics problems which arise in the genetic engineering field (transgenic research) are also notorious for consuming copious computer resources. A 96 picosecond simulation of a 1453 atom complex requires 960 hours (40 days) -- on a 1 MIPS platform -- before a complete solution of its trajectory is generated [Til88]. If one ponders the fact that the reverse transcriptase of the Human Immunodeficiency Virus (HIV) contains approximately 50 to 100 times as many atoms, the resolution of even a 100 picosecond trajectory becomes nightmarish and intractable on the fastest sequential platforms.

But the monetary and medical potential for genetically engineered drugs -- designer drugs -- is immense. Simulating the pharmacological and pharmacodynamic effects of an experimental pharmaceutical, rather than testing for adverse reactions in ten thousand mice, can save substantial resources. The effort to employ computer simulation for biotechnology is growing, since molecular biologists, like software engineers, are expensive to hire and maintain.

Molecular dynamics and fluid dynamics problems form part of the Grand Challenges of Computing [Rob89], 20 distinct areas of research deemed necessary for the collective economic security (survival) of the United States, as defined by the Defense Advanced Research Projects Agency (DARPA). Other nations, like Japan, and international consortia, like the European Economic Community, have recognized a similar list of problems deemed essential to their long-term economic viability. Table

---

[9] In principal, the savings generated from the multicomputer engineering process would be passed on to the consumer.

[10] Dozens of shared-memory multiprocessor cpu-hours are needed to complete the thermal and material transport solutions.

1.1 lists them, along with a brief description. Multicomputer technology will play a pivotal role in the exploration and resolution of these Challenges.

Approximately five of these 20 Challenges can be classified as real-time applications. One does not ordinarily think of weather/climate prediction as a real-time application. However, a 24 hour weather forecast is moot if it takes more than 24 hours to complete. And a 30 day forecast is meaningless if more than this interval elapses during the computation. A 24 hour-long real-time computation is an extreme example of a strict deadline (see Chapter 4).

Table 1.1 The Grand Challenges.

Science/Industry	Description
Combustion efficiency	Study of the quantum chemistry involved in combustion engines to create more efficient engines
Oil and gas recovery	Location of oil reserves and devising economic ways to extract them using seismic analysis and other techniques
Oceanic studies	Development of a global model of the ocean to help predict weather and other climate conditions
Speech	Computers than can interpret spoken languages
Vision	Development of machines and robots with improved vision capabilities for use in manufacturing and other applications
Antisubmarine warfare	Analysis of sonar and acoustic information to help track a new generation of quiet Soviet submarines
Nuclear fusion	Understanding the behavior of gases under magnetic fields to help develop fusion as a power source
Quantum chromodynamics	Calculations of the interactions of subatomic particles, used in predicting the creation of stars, new phases of matter and other phenomena
Astronomy	Analysis of data from radio telescopes to gain greater understanding of space, stars, and interplanetary systems
Transportation	Modeling new aircraft, space vehicles, ship and submarine hulls, and engine efficiency

Table 1.1 The Grand Challenges (cont'd).

Science/Industry	Description
Vehicle signature	Design of vehicles, aircraft and ships with reduced acoustic, electromagnetic and thermal signatures to avoid detection
Turbulence	Understanding the mechanics of turbulence in gases and liquids to help design vehicles
Vehicle dynamics	Analysis of the stability and ride of land, sea, and air vehicles
Human genome	Catalogue of all the gene sequences in the human being to understand the molecular and genetic basis of disease
Weather and climate prediction	Detailed atmospheric and ocean simulations to better understand ozone depletion, the dispersion of greenhouse gases in the atmosphere, climatic change; weather predictions in support of military missions
Materials science	Study of the atomic structure of specialty materials such as Si and GaAs, used in semiconductors
Semiconductor design	Analysis of how materials used in electronics operate, change their characteristics, and are structured
Superconductivity	Fundamental research into the properties of materials that conduct electricity with ultra-high efficiency
Structural biology	Analysis and visualization of three-dimensional structures of molecules that make up proteins, antibodies, and other cellular building blocks
Drug design	Assessment of interactions between molecules, chemical entities, and receptors based on computer simulation

Extraordinary computational resources are necessary to effectively implement a human speech recognition system. Phoneme recognition and other language-specific patterns can not be processed in a continuous and sustained fashion without substantial computation. Speech recognition systems must keep pace with normal human conversation. Currently, about one word per second is all that can be accomplished with a 1000-word vocabulary. The speech-driven typewriter of fiction will become a reality when the speed and reliability of verbal communication processing improves.

Vehicle dynamics stability calculations have an intrinsically real-time nature for safety and performance reasons. Modern jet aircraft, like the B-2 or F-16, are intrinsically unstable; dynamics calculations must occur at rates greater than 30 hertz, with 60 to 100 Hz routinely desired.

Vision and image processing are necessities for autonomous machines. A vehicle that can navigate over unfamiliar terrain must possess the skill to detect, recognize, and circumvent obstacles. The speed with which a vehicle's processing system can perform these tasks naturally determines the daily travel distance. The Mars Rover prototypes currently lumbering around laboratories move less than 6 inches each second. These ambulatory speeds must be reconciled with the exploratory goals of the planetary mission. The energy consumption of the Rover limits the total travel distance during the mission lifetime. If the processing rates that currently support a 6 inch/second velocity could be doubled, terrain exploration could be accomplished that much faster.

The global climate research challenge is estimated to generate spectacular amounts of raw data. The following quote is attributed to D. Allan Bromley, the science advisor to President George Bush [Mye92]: "'When the Mission to Planet Earth [project] sends up the Earth-Observing System A, it will send back every 4.8 days information equivalent to the entire contents of the Library of Congress,' Bromley said, giving another example of the need by supercomputers. 'Unless we have a factor of 100 to 1,000 in computer speed and capacity and similar factors for data transmission rates, we will simply be swamped -- unable to analyze the data and unable to get the information we need to make policy decisions in any reasonable time.'" The Library of Congress is estimated to contain approximately 25 terabytes of data.

In antisubmarine warfare, sonar and acoustic signature detection equipment serves as the eyes of an undersea vessel. Under hostile conditions, the rapid discrimination of friend from foe is a primary determinant for a vessel's survival. Processing acoustic signatures at sea is approximately equivalent to the speech recognition problem. A unique signature is continuously presented to a computer which searches a database for a known reference. The outcome provides evidence of an identifiable signal, of friendly or hostile origin, directing the ship's command authority to act. The faster that an acoustic signal can be analyzed, the more time available for planning, countermeasures, and consultation.

With the exception of the Challenges oriented at military applications, investigations aimed at the resolution and discovery of the underlying phenomena in each will result in tremendous societal gain and benefit. The creation of new industry, spawned from discoveries originating with the Challenges, will generate new growth and revitalize economies. A technological-industrial complex will supplant the existing military-industrial complex.

The Challenges point out the need for new simulations of a scalable nature. Scalable simulations run on multicomputers, and can effectively make use of the platform's processors. Building the scalable simulations necessary to tackle the Challenges, and the intrinsic complexity arising in each of them, will provide the opportunity for many engineers to participate in their construction. The intrinsic complexity found in the Challenges will not be elucidated or effectively explored via any sequential or shared-memory multiprocessor computing system; multicomputers of the SIMD and MIMD class are needed to expeditiously provide answers and explanations for phenomena that cannot be found by any other means.

1.3 Physical System Requirements for Multicomputers

How much computational resource does it take to address the Challenges? Can the necessary machine resources be economically packaged and will the machine run reliably? To answer the first of these questions, Figure 1.3 should be examined to gain an appreciation of the computational magnitude that the Challenges embrace.

The machine resources, in terms of memory size and floating-point bandwidth, are stated for each of the Challenges in Figure 1.3. Considering a word to be 64 bits, the majority of the problems in Figure 1.3 require memory subsystems physically addressable to 80 Gbytes. Supercomputers of the shared-memory multiprocessor class are currently available with 1 to 2 Gbytes; a few multicomputers are available with about 4 times this amount. The memory subsystems for multicomputers must either be equipped with 10 times more memory chips, or the individual RAM chips must shrink by this factor.[11] The 16 Mbit DRAM will soon be available, and this commodity will provide a factor of 4 expansion. The 128 Mbit DRAM is necessary before the 80 Gbyte memory subsystem becomes feasible to reliably manufacture and operate. VLSI components experience reductions in failure rates that mirror the continued rise in memory density and microprocessor speed increases.

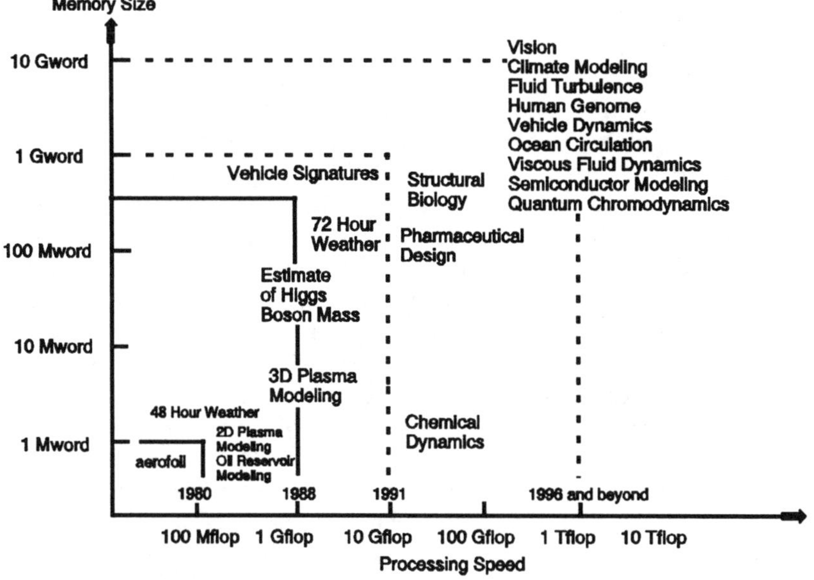

Figure 1.3 Machine resource requirements for the Challenges [FHP89].

The floating-point performance requirements trend toward 1 Tflop (1 trillion

---

[11] For an excellent discussion of VLSI feature scaling, system performance, and parallel processing, the reader is encouraged to see Seitz [Sei90].

floating-point operations per second). This is a terrific figure![12] The fastest benchmark results quoted for any computer system indicate that 6-8 Gflop sustained are possible for certain applications.[13] An increase by a factor of 120 is needed to reach the 1 Tflop region. Processor performance continues to scale with VLSI manufacturing processes. But memory subsystem bandwidth between the processors is as equally an impenetrable bottleneck for microprocessor-based multicomputers as it is shared-memory systems. A central barrier for obtaining a 1 Tflops computer system is the memory subsystem cycle time.

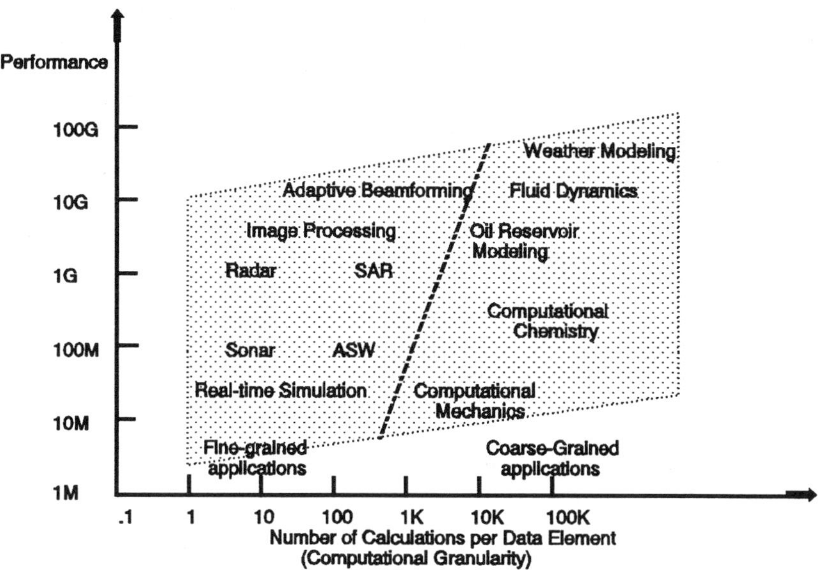

Figure 1.4 Granularity classification of the Challenges. (Courtesy of Intel Supercomputer Systems Division. Copyright © 1991. All rights reserved.)

The fastest shared-memory systems are capable of a 2 or 3 nanosecond pipelined memory fetch cycle (from registers, not main memory). Attempting to replicate this level of performance on a distributed-memory multicomputer may be ineffective and uneconomical, as message-passing operations compete with the traditional load/store memory functions. To reach 1 Tflops on a 1000 processor multicomputer also implies that each must sustain 1 Gflops. General purpose microprocessors are routinely capable of about 20 Mflops (2% of the desired 1

[12] The United States National Debt is also measured in trillions; it stands at about US $4 trillion.

[13] A distributed memory SIMD-class computer holds this record (an extreme form of chest pounding).

Gflops). A 1 Gflops sustained performance by a microprocessor will require 80 MHz clocks, 128 bit words, 0.1μ CMOS feature sizes, or a combination/compromise of these three. Whether such clock speeds and interconnect densities are possible to economically manufacture in CMOS VLSI remains to be seen.

Figure 1.4 illustrates the granularity classification of the Challenges. The figure shows that the quantity of computations per data element -- the computational complexity -- varies substantially with each Challenge. At one end, certain problems in image processing require a literal handful of arithmetic operations to complete. These are fine-grained applications; each process possesses a small number of arithmetic computations, and is easy to construct. At the other, weather computations consume thousands. These are coarse-grained applications; each process is composed of complicated arithmetic operations and mathematical manipulations. Casting a general circulation model into a scalable context can be a tricky and complex effort (see [Bou77] to appreciate the complexity of atmospheric flow modeling).

The computational complexity found in the Challenges spans many orders of magnitude. The disparity accents the requirement to construct many different classes of multicomputer system. Some systems need not be equipped with the biggest word size, or the greatest floating-point performance, while other manifestations must be technologically superior in every respect.

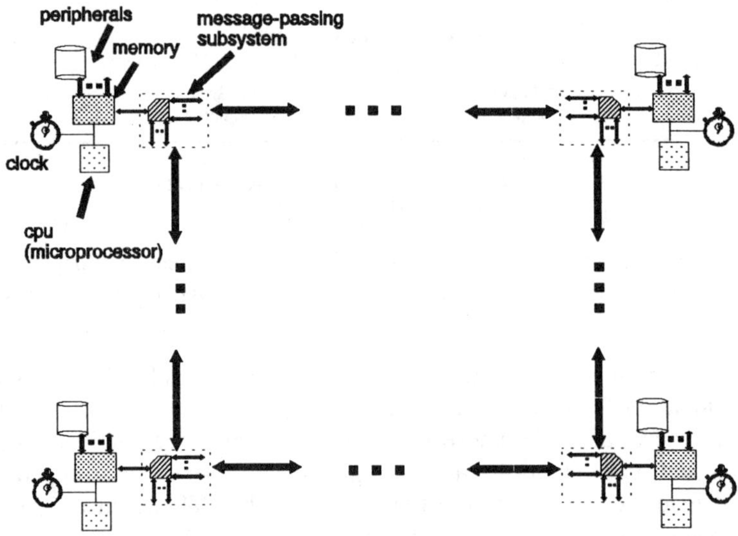

Figure 1.5 A scalable multicomputer architecture.

An equally important requirement for a multicomputer is the I/O support. Moving information from disk to memory in a distributed system is more costly than the equivalent operation in a shared-memory multiprocessor. The exception for a multicomputer architecture is illustrated in Figure 1.5. This totally scalable multicomputer architecture possesses I/O peripherals and computational support which are not shared, except through message-passing primitives. Message-passing primitives replace the physical interconnection found in shared-memory systems. Access to a resource within the multicomputer is on per-request basis, rather than by default.

Each node in the Figure 1.5 architecture schematic could just as easily be a low-end personal computer with cheap hard disk, or a custom computer system. A multicomputer can be packaged in many physical configurations. The hardware that comprises them has acquired a commodity-like status, but the software determines how effective the system can be. The software enables the multicomputer to perform work, substituting (simulating) the equivalent of many humans. The multicomputer's potential to execute scalable simulations advances the goals of efficiency in the production of goods, services, and discovery.

Concluding Remarks

While multicomputer systems offer the potential to explore complicated and advanced problems crucial to the survival of mankind, the technology can also be exploited and misused. The eminent power manifest in multicomputer systems resides in the software built for a specific purpose. The growing societal dependence on computers will invite insidious individuals and corrupt corporate entities to leverage scalable computation into unfair and potentially pernicious enterprises [Pag88]. These incidents will be difficult to prevent, and nearly impossible to undo.

A teraflop computer, like a nuclear weapon, projects a strategic and metaphorical threat. The power contained within each edifice can be highly destructive. While the nuclear weapon can physically destroy a city, state, or nation, a teraflop computer can raise havoc with electronic infrastructure and confound an ill-equipped bureaucracy. Vast and rapid shifts in monetary and economic fortunes can be implemented with a multicomputer of sufficient power. Financially strategic interests can be plundered and extracted leaving behind destructive consequences reminiscent of a nuclear detonation. The seeds of a potentially devastating tragedy are likely being already sown [Mar91].[14] The author's opinions on the dangers of intentionally criminal software are stated elsewhere [Ste92a].

Suggested Reading

An excellent overview of parallel computing appeared in the October, 1987 issue of *Scientific American*. The classic reference article to multicomputer systems is C.L. Seitz's [Sei85] publication on the "Cosmic Cube." An early *Physics Today* paper by Fox and Otto [Fox84] categorizes many physics problems ripe for multicomputer solution. The *Physics Today* paper by Bowler *et al.* [Bow87] sites numerous examples of large-scale transputer-based systems for theoretical physics. Some of the earliest

---

[14] One of the biggest ironies of the 20th century lies in the fact that the very technology we create to "improve our existence" often has the same potential to instantaneously make it infinitely more miserable.

demonstrations of multicomputer problem solving effectiveness can be found in the Hypercube Multiprocessor conferences ([Hyp87] and [Hyp88]). A general reference text on parallel processor hardware and software, including multicomputers, is that of Almasi and Gottlieb [Alm89]. For a more rigorous discussion of efficiency and performance analysis for parallel computer systems, see [Fox88a] (especially sections 3.4-7). Refer to Braunl's Law [Brau91] for an extension of Amdahl's Law as it applies to SIMD and MIMD computers. The Association for Computing Machinery's journal *Computers and Society* is an excellent quarterly publication on the sociological impact of computers.

2

Project Planning and Preparation

This chapter introduces the proposal writing process as a vehicle to obtain the necessary funding and approval to develop a multicomputer product. Proposal writing, like all technical document preparation, and composition in general, is a skill unto itself where clarity of discussion and brevity of conveyance are valued commodities. The author believes that successful projects are guided by informed and articulated judgement predisposed to planning and organizing detailed resource execution.

A detailed project plan embodies these attributes. Writing an effective proposal which receives funding is a tedious process. Many hours of labor must be dedicated to composition, and extensive revision to refine an idea is mandatory. But the time invested during the proposal phase can have a huge payoff during the later phases of a funded project. The payoff can be realized from several sources.

The estimates that accompany the proposed project may in fact be correct, and the team engaged for the project will complete it on time, and within budget. This is a terrifically rewarding event, however rare in practice it may be. Alternatively, the proposed project may have outlined several technological issues to resolve, each keyed to a specific timetable, but ultimately prove impossible to conduct under the proposed schedule. Risk management is essential for complicated projects. Where possible, a proposal should categorize these risks.

The key to successful cost estimation is the articulation of a solid and concise project plan. The better the plan, the less likely that "surprises" will emerge during a crucial phase. Unless the organization undertaking the development of a specific project has already sufficient history, be assured that "surprises" which threaten nominal execution will most certainly arise. Even organizations with substantial experience building systems, such as the automobile industry, are almost notorious for schedule slippage and cost overruns.

No cookbook formula or tools can eliminate uncertainty from a development effort. About the best an organization can do is to manage change, and minimize the uncertainty arising from it. Successful organizations are precisely equipped to do this; unsuccessful ones fail at the attempt.

2.1 Project Planning Basics

The project plan integrates many elements. A project plan enumerates specific goals and their associated costs, and includes a method for executing the individual tasks in a controlled manner.

An accurate cost estimate hinges on the organization of an effective project plan. The project plan serves as the principal task definition document. It describes the project execution sequence of the scheduled tasks, states the desired objectives, and enumerates deliverable quantities. A project plan is comprised of several major sections. These include statements of deliverable items, innovative claims and advantages of the project, technology transfer, technology description, discussion, and comparison, resource allocation -- which may include a statement of work (SOW) and work breakdown structure, a detailed schedule, resumés of principal investigators and participants, marketing research, literature references, and a facility description for starters!

Each item is incorporated into the project plan, which may form the basis of a proposal for submission to management or a funding agency. Many companies and funding agencies have preordained proposal formats.

With the exception of marketing research, and the work breakdown structure (WBS), all of the previously mentioned elements appear in proposal formats requested by the Defense Advanced Research Project Agency (DARPA). This format is prototypical for all U.S. government funding agencies. The DARPA proposal typically does not exceed 50 pages (appendices excepted).

The author has written many proposals -- not all of them successful -- to several agencies requesting funds to investigate software for parallel computation systems. Effective proposal writing is a valuable skill, but is not in itself enough to guarantee a grant; you must be lucky as well. Nevertheless, it is helpful to outline the content and structure of a proposal as a means to develop familiarity with project plan design and specification.

2.1.1 The Idea: Foundation for the Project Plan

An idea often motivates the construction of a project plan. Ideas are not necessarily cheap or commonplace. Carefully articulated ideas proposing innovation, cost-effectiveness, or even those ushering new, basic technology into the marketplace require a plan to develop. Often the idea is highly abstract, and initially only understood by the inventor. It is his or her job to elucidate the nuances to an intelligent audience, and demonstrate the novelty and potential.

No matter how clever or earth-shattering, if the idea or concept is beyond explanation, agency funding is not likely to follow. Grandfathering a sophisticated product to market mandates stamina, desire, and patience. Sometimes, Machiavellian circumstances stifle honest proposal efforts, and destroy all desire -- and hope -- to obtain domestic funding [Dav91]. Many regulatory, bureaucratic, and organizational roadblocks lie in wait to stifle product development. Anticipating these hurdles at the earliest possible stage of product development is a strong reason for creating the project plan.

2.1.2 Statement of Innovative Claims

A declaration of uniqueness, the *statement of innovative claims*, is found among the first pages of the DARPA proposal. The innovative claims summary is typically a one page capsulized digest or compendium of the central goals and objectives that the project attempts to accomplish. In many ways, this statement is the resumé of creativity, ingenuity, and insight into the engineering of your idea. The following example outlines the substance and style of an innovative claims statement:

> This investigation seeks to establish the first instance of a unique display subsystem, the tessellated display, as an instance of a MIMD-class, real-time multicomputer specifically organized to support the visualization of gigabyte- or terabyte-sized datasets. This scalable concurrent visualization system (SCVS) (see [Ste92c]), and its accompanying visualization environment, the Overseer, will also demonstrate the temporal coordination of an MIMD-class multicomputer system in real-time by applying a unique global time standard known as local area multicomputer time (LAMT).
>
> The Overseer environment incorporates a systolic compression and expansion (SCE) load balancing mechanism, an approach which is believed to facilitate high-interactivity (e.g., real-time) display and rendering of graphical databases, a first for tessellated display subsystems.
>
> The Overseer window interface supports multiple simultaneous users of the SCVS platform, a networked component of a local area multicomputer (LAM) system [Bur89]. Additionally, Overseer provides a mechanism to seamlessly integrate the tessellated display elements into a coherent and modular display ensemble.
>
> A preliminary performance analysis (based on current silicon process technology) of the SCVS indicates that a one million polygon database may be transformed, clipped, shaded and rendered at an interactive rate greater than 20 Hz, without graphics hardware accelerators. The displayable area for this 32 by 32 tessellation composed of 512 by 512 pixel display elements is 16384 by 16348 pixels.

When examined by a knowledgeable referee, the innovative claims statement distinguishes your idea from a competitor's, and introduces the detail of the subject in a concise and meaningful way. Restricting the length of this statement to less than one page forces a content focus on the idea's essence; and is also a DARPA proposal format requirement.

2.1.3 Deliverable Item Summary

The idea you propose to investigate or develop must produce corporeal results during project execution and at the conclusion. Whether the result is documentation, tape recordings, videotapes, or computer software/hardware, the *deliverable item summary* states the expected results and items to be generated during the course of conducting research and development. When the research is concluded, specific items will be made available for inspection or be physically deliverable to the funding agency as proof of execution, whether it was successful or otherwise. An example of this summary follows:

> We propose a two phase effort. Generally, the deliverables from Phase I are draft documentation and proof of feasibility models. The deliverables for Phase II are final documentation and a working SCVS integrated with an alpha Release Overseer. Additionally, we propose to deliver monthly Activity Reports that will contain schedule, activity, and budgetary data summaries.
>
> **Phase I.** We propose to deliver the following items at the completion of the Phase I effort: (1) a draft system specification defining what Overseer is and what it will do, (2) a draft user's guide defining how to use Overseer, (3) a draft software requirements specification defining what the software must do to permit Overseer functionality, (4) a limited capability, 3 by 3 tessellated display hardware prototype based on a UNIX workstation host.
>
> **Phase II.** We propose to deliver the following items at the completion of the Phase II effort: (1) a set of final system documentation. This includes a system specification, user's guide, software requirements specification, manufacturer's supplied documentation for off-the-shelf items, and a complete set of hardware design and fabrication drawings, (2) a training guide that includes a videotape showing actual use of Overseer in operational and development applications with SCE load balancing, including the LAMT global synchronization system, (3) an Overseer environment consisting of a host UNIX workstation, and an 8 by 8 tessellated display. Please note that the prototype hardware delivered in Phase I will be upgraded to be included into the Phase II unit.

This example deliverable item summary is organized according to a phased development effort. Each phase of the development has specific deliverable items associated with the investigation. The monthly activity summaries, a project management tool for cost and project performance monitoring, are shared with the client -- the government in this case. In a commercial corporation, these activity summaries may be handled by the cost accounting department which records personnel expenditures and project costs.

The principal engineering deliverables during Phase I include draft documentation and design information. That preliminary information generated and organized in Phase I illustrates the prototypical nature of the project. Had this project been an adjunct to a very mature technology, such as machining technology for fasteners, then the Phase I design stage would probably circumscribe the extent of the entire investigation.

2.1.3.1 Risk Assessment and Rapid Prototyping

The relationship between deliverables and the project phase is important where technological or development risk is apparent. Projects which attempt to "push the envelope" of scientific knowledge or development have large risk factors. These factors are varied, often unpredictable, and project-specific. Consider a project to develop a high-T_c superconductor. How will a materials engineer determine the precise stoichiometry of the elements to blend? Since no formula or tool readily exists to predict the relationship between stoichiometry and critical temperature, a costly trial-and-error period of experimentation is needed.

To prevent costs from sky-rocketing, the experiment must be parameterized. A restricted combination of stoichiometries will be permitted during early research phases, and other possibilities reserved for later phases.

This methodology, termed "rapid prototyping," is a viable approach to risk management. Rapid prototyping permits the execution of a scaled-down effort, one which incorporates the barest essentials of the idea. Prototyping is really a form of simulation, where the characteristics of the full-up system are emulated by another, less-costly alternative for concept verification.

Rapid prototyping occurs in many forms. The Very High Speed Integrated Circuit Definition Language (VHDL) [IEEE1076] permits complete digital circuits to be simulated at the gate level prior to fabrication and layout. A simulation of this system is far more cost-effective than a full-up development effort, which may not work at all when complete.

Some organizations are well suited to executing rapid prototypes. These include small firms with low overhead, or profit centers within large corporations who can operate in a less restricted manner than their corporate peers.

A deliverable item is at risk, from the perspective of implementing a new computer system, when either the technology is largely undeveloped, as in the cases of a real-time multicomputer or fault-tolerant massively parallel processing systems, or if the user community is undereducated as to how a technology can be applied to solve their problem. For parallel processing, both the intrinsic technology and user education issues are risk items.

A *technical* risk item is one that faces the possibility of not working correctly because the technology could not be engineered into a functional unit. The technology has not been sufficiently tamed to permit the rapid prototype to operate according to design. In the case of high-T_c superconductors, the stoichiometry is a technical risk factor.

A *social* risk item arises when a technology can be engineered to solve a problem, but the skill level of the user base is not sufficient to comprehend the operational characteristics. In parallel processing, few organizations have trained individuals to construct load balances for multicomputers. This educational deficiency constrains the widespread acceptance of parallel processing systems.

Rapid prototyping methodology provides the mechanism to construct or simulate risk items which are difficult or costly to specify. Real-time multicomputers, computation systems coordinated in a temporally synchronous fashion, can be partially simulated, and methods to educate a user community for this technology can be outlined and developed.

One should note that rapid prototyping, a vehicle to quickly test the viability of a hypothesis, is not directly supported by current U.S. government procurement practices. This restriction burdens the proposal writer with developing a full life-cycle development model for the proposed research effort. Requirement specification through design, integration, testing, and maintenance efforts is difficult to organize when "pushing the envelope" with new technology.

Alternatively, large-scale engineering efforts, such as those required for the reinvigoration of the U.S. air-traffic control system or for the proposed Earth Observing System Data and Information System (EOSDIS) require extraordinary resources and time to carry out. Among many requirements, EOSDIS must possess the capability to archive, process, classify, and visualize at least one terabyte of raw data each day for over five years. Specifying the computational processing system to be a serial variety would be foolish. But what kind of parallel technology should be used? What is the best way to phase in a technology, and then transparently upgrade it with newer, faster technology without halting the data stream?

Harnessing rapid prototype methodology is the most likely way to implement an EOSDIS which can evolve with parallel processing technology.

2.1.4 Cost and Schedule Summary

A quick method to express the development time frame and cost resource requirements for the project is found in the *cost and schedule summary*. This summary is typically no longer than one page, and merely enumerates the principle program or technical objectives, their associated costs, and the time to complete them. The summary is useful to convey cost and schedule issues to those individuals who have little time to examine the proposal contents.

The schedule and cost summary is ideal for executives who must delegate technical decision making authority, and require their delegates to notify them if the proposal contains sufficient technical merit and innovative strength. The chain of command in many businesses and organizations operates in this fashion. The technocrats pass their judgement to the executives who allocate budget and resources for projects. If the project is technically meritorious, and does not exceed the available budgetary resources, then it may be funded. Table 2.1 illustrates a cost summary for the SCVS project described in the following sections.

A brief description of the cost summary detail items is given. Each item is identified by name, and a corresponding statement of the tasks purpose will also appear in the statement of work. The expected duration of each task is also provided. This estimate appears calendar months. The labor cost associated with each task is stated. It is the product of the calendar time, the number of persons assigned to the task, and an appropriate labor rate. If any hardware or software items much be procured to complete a task, the cost of these items is stated. Any travel needed for meetings or customer interface in support of each task is listed. If an expert or consultant is needed to complete a given task, the cost for this person's time is stated. General and admistrative overhead (G&A), and profits for the task are also listed.

	Task	Time[1]	Labor[2]	H/W & S/W	Travel	Experts	G&A	Profit	Total Cost
PhaseI									
First Year									
I.1	SCVS Plan	1.0	5.3			5.0			
I.2	Define Sys. Reqmnts.	1.5	10.6			10.0			
I.3	Define Overseer S/W Reqmnts.	1.5	10.6	10.0		5.0			
I.4	Define SCVS H/W Reqmnts.	2.0	15.9		5.0	5.0			
I.5	Develop & Exec. Overseer Sim.	4.0	85.2	41.0		5.0			
I.6	Build 3x3 SCVS Proto.	4.0	63.9	67.5	5.0	5.0			
Second Year[3]									
I.7	Proto.Int.	4.0	66.8			5.0			
I.8	Prepare Draft User Man.	1.5	22.3			3.0			
Ph.I Subtotal		16.0	280.6	118.5	10.0	43.0	27.0	38.3	517.5
PhaseII									
II.1	Finalize Sys. Specs.	2.0	33.4			10.0			
II.2	Build 8x8 SCVS	5.0	133.5	287.0	5.0	15.0			
Third Year[3]									
II.3	Develop Overseer S/W	6.0	174.4			15.0			
II.4	System Int.	5.0	139.5			20.0			
II.5	System Test	2.0	46.5		5.0	5.0			
II.6	Doc. & Training Matl.	3.0	69.8		5.0	10.0			

	Task	Time[1]	Labor[2]	H/W &S/W	Travel	Experts	G&A	Profit	Total Cost
Ph.II Subtotals		20.0	597.1	287.0	15.0	75.0	58.4	82.6	1,115.1
Totals Phase I & II		36	877.7	405.5	25.0	118.0	85.5	120.9	1,622.7

1: task calendar time overlap not shown; 2: includes 70% labor overhead; 3: adjusted @ 4.5%/yr inflation. Time in calendar-months. Cost in US $1000.

The cost and schedule summary may be used as a quick check of resource allocation. For instance, if you estimate that three full-time engineers are expected to work on a project for eight months, then the total cost for their time should be easily visible in this summary. Thus, if an engineer earns US $5000/month, over eight months this amounts to US $40000 to support his or her efforts.

Material costs are bundled with the cost and schedule summary. The cost summary may detail individual material costs (such as computers, software, electronic components, etc.). Major tasks are stated, and each is given a cost. Each phase of the project is assigned a total duration. Each cost item in the summary is keyed to an activity found in the statement of work. This concordance is useful for program monitoring, where budgetary concerns are matched against program objectives.

The summary is organized to account for a phased development effort. During each phase, the costs associated with a specific task, objective, or milestone are stated, and the execution time for the entire phase is stated. This itemized breakdown should roughly correspond with the deliverable item summary, and be traceable to the innovative claims summary as well.

2.1.5 Statement of Work

The specific tasks, project scope, and requirements to be fulfilled by the investigation are often located in the *statement of work*. This portion of the plan is often the most important, from a contractual point of view, as the exacting nature of the work to be performed is detailed and described here. The statement of work (SOW) must clearly enumerate the nature of all tasks required to complete to the project, and develop the proposed idea. These tasks form the heart of the investigation.

Clear descriptions of each task, and their relationship to others, define and strengthen the project organization and conduct of the investigation. SOW preparation is an art form unto itself, and is best conducted by a senior program manager and/or hardened engineer. Drafting an SOW solo is the mark of a capable systems engineer. A neophyte engineer graduates to the systems engineering world when he or she can draft an SOW.

In the simplest context, an SOW enumerates the scope of the work needed to achieve the deliverables (section 2.1.3) and sucessfully reach the projected milestones (section 2.1.4). If project management and cost reporting is required, then the SOW should say this. If paperclip counting is required thrice daily, then the SOW should state this also. Each task to be performed during the execution of the investigation should have an associated SOW paragraph or item.

Imagine what the SOW must look like for the space shuttle, the B1-B bomber, an underground subway, or other large-scale project. These projects have SOWs that number in the hundreds or thousands of pages. Be assured that each item will have an associated audit function to track and verify whether the contractor (the engineering team) is correctly and legally performing the effort. A small portion of a SOW is listed below:

1.0 Phases

We propose a two-phase program. The first phase, to be completed in 18 months, provides DARPA with a complete but preliminary definition of the SCVS and Overseer visualization environment. During the second phase, we will formalize Overseer documentation and develop a working demonstration system.

During Phase I, we will develop a draft system specification, a draft software requirements specification, and a draft user's guide. In addition, we will procure sufficient hardware items in order to build prototypes of the SCVS with tessellated display item and software prototypes of, as a minimum, portions of the Overseer's SCE load balancing and LAMT maintenance algorithms.

Phase II has an estimated 24 month duration, and the above mentioned documentation will be finalized, the SCVS tessellated display subsystem designed and finalized, a training video designed and completed, the required software developed, and the Overseer system integrated and checked-out.

In addition, program planning and control will constitute an on-going effort throughout both program phases.

1.1 Program Planning

Immediately upon contract award we will develop detailed task definitions, detailed project schedules and budgets, track project performance against the schedules and budgets, and periodically document project performance in activity reports. In addition, an element of the program planning and control task will be to formally submit all deliverables to DARPA.

We feel that any project must have a plan, no matter how small or revolutionary its objectives. We will define all the tasks necessary for both phases of the program. Span times will be developed based upon historic data recorded from similar efforts. A detailed program schedule will then be developed based upon efficient work flow, resource availability and span times.

These data will be documented and provided to DARPA with the first monthly Activity Report. Program planning and control will be the single point used by our organization for delivery of contractual items.

Phase I Tasks and Activities

1.2 Develop Draft System Specifications

This is the first key to success. This effort emphasizes defining the context and purpose for Overseer, identifying the major functional components, and how they work together. SCVS and Overseer environment objectives will be stated in the detail required to allow for system development and test. System architecture and performance studies will be executed. Formal methods of system specification, such as that afforded by the Z specification language, will be used where applicable.

1.3 Develop Draft Overseer Software Requirements Specification

System requirements will be documented as they evolve. A preliminary test philosophy will be developed at this time. The test philosophy will aid in establishing the exact statement of requirements. Two technical innovations contained in this proposal are the development of a tessellated display, and the development of the Overseer environment and concurrent visualization system engine that equalizes computational load and provides seamless display connectivity in real-time. Software rapid prototypes of the concurrent visualization system engine will be developed early on to provide proof of concept via simulation. Specification development will proceed via formal methods using Z and supporting computer-aided software engineering (CASE) tools to maintain DoD-Std-2167 documentation.

1.4 Define SCVS Hardware Requirements

This task is divided between the two phases. A UNIX workstation and items to develop a tessellated display prototype will be procured during Phase I. Items to complete the display and its interface will be procured during Phase 2. Procurement will be to "fly sheet" specifications. Hardware specifications for SCVS instances and display elements, packaging alternatives, and interface definitions will be prepared.

1.5 Develop Draft User's Guide

A user's guide will be prepared in the early stages of Overseer development to ensure the utility of the system. This effort will be accomplished in parallel with the Overseer user interface definition taking place in the system specification development task.

1.6 Develop Tessellated Display and Overseer Prototypes

A prototype of the tessellated display, in a 3 by 3 format, will be developed early on to provide proof of concept. This effort will include exploring packaging details as well as functional performance. All work will be accomplished in accordance with engineering sketches developed in this task. The SCVS and tessellated display prototype will be denoted as *Raptor E.*

The Overseer prototype development will explore the scalability issue for rendering with ray-tracing algorithms and geometric (polygon) entities. Overhead associated with message-passing, global synchronization, user interface interaction, and load balancing will be assessed and compared against theoretical margins.

Phase 2 Tasks and Activities

2.1 Develop Overseer Software

Overseer software will be developed in a tailored DoD-Std-2167 environment. It is our intent to allow for full visibility into the software without the constricting efforts of full configuration control. Software requirements, design and implementation will be fully documented. To facilitate documentation and traceability, CASE techniques will be applied to the software development effort. An appropriate CASE tool (such as a formal specification language processor) will be adopted for this purpose. Development and test history will be kept in engineering notebooks maintained by project personnel. The LAMT synchronization algorithm, SCE load balancing, GUI port, graphics library port, and the DRSM will be developed in this environment. Software development will be in the "C" language. The UNIX workstation procured above will serve as the development and delivery environment.

2.2 Develop Tessellated Display

The tessellated display prototype developed in Phase I will be scaled to the 8 by 8 deliverable display. Display drawings will be updated to reflect this new configuration.

2.3 Integrate Overseer

This task is for the integration of the hardware elements into a functional entity and then, with the Overseer environment loaded, ensure that the system functions as specified in the system requirements.

2.4 Develop Training Video

It is our intent that this technology be broadcast throughout the user community. To this end, a training video will be developed that will show Overseer participation in the solution of several types of problems. This task includes the development of at least two distinct problems for solution on Overseer with SCE load balancing. It includes developing a video script and the development of actual training material.

2.5 Finalize Documentation

A set of final system documentation will be developed in this task. This includes a system specification, user's guide, software requirements specification, manufacturer-supplied documentation for off-the-shelf items, and a complete set of hardware design and fabrication drawings. These documents will be tailored to MIL-spec format.

Once again, the phased development approach is used to organize the SOW. Phased development helps to minimize technical and project risk by reducing the scope of specific tasks. Scaling the development effort into different phases provides time for project staff to articulate design and build prototypes or simulations.

Another prime motivation for the phased approach is to control development costs by segregating the project into smaller segments, each with a clearly defined objective. Big projects are often superpositions of smaller projects. By addressing the smaller projects in a controlled fashion, expense and risk is contained.

The statement of work summary is limited to three pages in the DARPA proposal format.

2.1.6 Technology Transfer

The technology transfer section discusses the potential benefits of the project, as the user would see it, once the technology is developed. Technology transfer describes how the resulting effort enables users to apply the developed system. Perhaps the technology will save thousands of lives, or provide unlimited electrical power at no cost. Whatever the project promises to create or deliver, the technology transfer will illustrate how, for example, the results generate an integrated circuit manufacturing process incorporating high-T_c superconductors, offer the user a quicker method for finding prime numbers, construct a focal-plane image compression computation system, or articulate a linear scalable visualization system.

The goals stated in this section summarize the principle technological merit of the proposed effort. An example of a technology transfer statement is:

Technology Transfer

The development effort proposed for this investigation will lead to the Overseer environment: a rendering system, scalable graphics library, and associated utilities for visualizing data created from scalable concurrent visualization systems (SCVS). SCVS are contentionless graphics platforms. They are organized as MIMD-class multicomputers with each node possessing a geometry transformation system, small frame-buffer, and display device dedicated to the generation of pixel-based information in real-time. The Overseer environment is an integral part of the SCVS, as it supports the display list generation, graphical primitive interface, user interface, load balancing, global synchronization, and graphical output for images. The technology transfer and principle goals for the SCVS engineering effort are as follows.

1. Scalable Software Metric Construction

The Overseer, or a product of similar scope and functionality, is a mandatory requirement to permit the eventual construction of SCVS. Once the software has been developed, the SCVS hardware will be enabled to achieve its performance objectives. Because engineering scalable computation software is a unique activity, and one which does not yet have a software metric, very valuable information pertaining to scalable software metrics will be discerned by developing the Overseer software under a DoD-2167 standard environment. This information, in the form of published reports and documents, will assist similar efforts in the future.

2. Marketplace Gap

The Earth Observing System (EOS) and Superconducting Supercollider (SSC) projects will likely produce 1 Tbyte datasets on a daily basis. How will these applications perform data reduction for these volumes in an expeditious manner? Shared-memory systems are contention-bound, and scale only with VLSI processes. The SCVS packages a scalable computation and visualization mechanism together, and thus offers a scalable packaging and computation architecture. These are the tools which will likely find roles in the solution to these advanced applications.

No workstation or parallel computer manufacturer offers a current solution which is scalable for both computation and visualization needs. Thus, the SCVS will establish a new alternative, based on entirely scalable software and hardware packaging technology, which may leapfrog the current generation of systems, as they do not demonstrate a scalability beyond the refinements in silicon process technology.

3. Real-time Multicomputer Systems Engineering

The notion of time is key to the execution of the Overseer environment, and to the SCVS real-time multicomputer in general. Real-time systems, and their engineering, are of tremendous importance to both military and civilian systems. As reliance on computers grows, the demand to expand into massively parallel computation systems will rise, and real-time multicomputers will also become prevalent. Therefore, it is essential to understand and develop techniques for specifying, designing, and integrating real-time multicomputer systems. Little of this work has been conducted to date.

The SCVS, a real-time multicomputer, depends on the Overseer for several critical functions, the most important being the keeping and updating of local area multicomputer time (LAMT). This quantity is used to synchronize the execution, processing, and updating of each tessellated display unit. The Overseer timekeeper is responsible for the global synchronization and maintenance of LAMT for each instance of the SCVS. It is essential that each instance of the SCVS display update occur within time τ_u of the global time reference. This prevents the emergence of a discernible twinkling, an incoherency resulting from the asynchronous update of the independent display units.

The establishment and maintenance of LAMT by the Overseer timekeeper, within the SCVS framework, adds visibility to the goal of real-time multicomputer contexts. Since the results of each computation can be seen on a frame-by-frame basis, the success or failure of specific methods and procedures for implementing LAMT will be immediately known. Overseer will be evaluated in the testbed hardware developed for this effort, providing quantitative proof of the success or failure of specific LAMT organization techniques.

The engineering effort undertaken during the development of Overseer will demonstrate methods for implementing LAMT in a real-time multicomputer environment. This knowledge, once published, will become significantly more valuable to organizations aspiring to create other real-time multicomputer applications.

4. Scalable Graphics Library Standard

The Overseer rendering system will be based on an existing public domain graphics library. This library will provide a mechanism for easily transporting applications and datasets between disparate host platforms. It will also serve as the foundation for the development of a scalable version of a window interface.

5. Tessellated Display Subsystem Window Interface

The Overseer environment will be based on windows, icons, and pull-down menus. It will be a scalable rendering environment. This implies that if only half of an SCVS platform is required, then the user need only reserve half of the computation resources, and render into half of the display area while another user, or many users, divide up the remaining computation and display subsystems. This is only possible through the scalable nature of Overseer, the SCVS tessellated display subsystem architecture, and the local area multicomputer concept (LAM). The LAM concept is discussed in the technical rationale section.

6. Debugging Real-time Multicomputer Systems

To assist with the integration and debugging effort for real-time multicomputer software engineering, a special tool, the distributed real-time symbolic monitor (DRSM) must be constructed. The DRSM permits a user to symbolically observe and modify remote address spaces in real-time. While break-point symbolic debuggers, such as dbx, enable the logical structure of a simulation to be analyzed, they destroy the temporal coherency of a real-time simulation, and cannot be used for real-time debugging.

Constructing the DRSM, and applying it to the engineering of Overseer, will validate the DRSM. It will support acceptance test procedure (ATP) execution, and supply vital statistics for load balancing and LAMT performance. The DRSM will be released with the Overseer product.

7. Load Balancing via Systolic Compression and Expansion

Obtaining a load balance for MIMD systems is a critical task. Any multicomputer application requires a load balance to achieve maximal speedup, since the most heavily laden computation element influences the effective execution speed. The two most popular techniques for load balancing, in terms of spatial decomposition, for MIMD applications development are either simulated annealing or "eye-balling." Simulated annealing is useful for sparse systems, or irregular geometries, while eye-balling is simple and quick but applies only to very regular and highly symmetric structures.

Systolic compression and expansion (SCE) load balancing is a mechanism which governs message-passing between adjacent SCVS instances. Each instance has a finite quantity of RAM, and this constrains an instance's capacity to accept geometry from neighboring nodes resulting from scene transformation or eyepoint translation. A "clipping force" determines the rate of data transmission between neighboring nodes. The memory high-water mark for display list entries will limit the addition of primitives to a node display list, and cause a suspension of message-passing for geometry insertion. All remaining Overseer message-passing functions, such as LAMT maintenance and user interface input will continue. We will develop this mechanism as a part of the Overseer.

This is the compression mode of SCE. It is active when a long-range view of the scene is needed. Attempting to examine a dataset from a long distance forces the object to become point-like in appearance, and this condition amounts to an attempt at rendering a very large dataset, composed of hundreds of megabytes (or more), on just a few SCVS instances which may possess only a few tens of megabytes each for display list processing.

On expansion, however, the viewpoint is more of a zoom or closeup of a small region. This produces a magnification of scene geometry. The greater the zoom, the more likely a few geometric primitives will replicated on many SCVS instances. SCE load balancing is discussed in section on Technical Rationale.

The items presented in the technology transfer section state engineering objectives. In this case, the proposed software and hardware system offers benefits not found among commercially available platforms. Of note in this example are the goals of exploiting logical concurrency in the software design process to organize a linear scalable software library, the notion of specifying and maintaining "local area multicomputer time" for the real-time synchronization of a multicomputer, the debugging tools for real-time multicomputer development, the contentionless display subsystem, and SCE load balancing.

The technology transfer summary is limited to five pages in the DARPA proposal format.

2.1.7 Technical Rationale

The *technical rationale* section may be the largest portion of the proposal. In this section, the author must describe the technical arguments needed to substantiate the innovative claims (recall section 2.1.2), a *technical approach* description which substantiates the deliverable item and cost/schedule summaries, and a *comparison* of the proposed effort to ongoing research in the field, and how the proposed effort will supersede it.

These three subsections are especially valuable for project planning and execution. The rationale portion outlines how the innovative claims will be realized from the conduct of the technical approach. The rationale provides supporting evidence, based on literature searches, or theoretical arguments which give heuristic justification -- a kind of existence proof -- for the likelihood that the proposed project will succeed. The rationale must demonstrate the possibility that a coherent investigation can approximately or actually satisfy and realize the innovative claims.

The technical approach outlines how the innovative claims will be reached, and this is the heart of the project plan. The statement of work details all the individual classes of activities to be performed, but the approach specifies the collective problem solving application of the individual classes. Thus, systems engineering tasks to enumerate the extent of hardware and software interfaces, or software design tasks to construct a real-time executive are both valid statements to be found within the approach subsection.

The effort invested to craft a complete project execution sequence in this section effectively serves as a simulation -- a *gedanken experiment* -- of the engineering activity for the project. Imagine placing problem solving skills into the various phases of the project. A participatory effort is engaged by the author who realizes, and may even "see," all the activities taking place.

This experience is clearly much easier to accomplish if a single author is responsible for the technical rationale. But very large projects, such as an air traffic control system, force a soliloquy to become a chorus of effort.

The comparison section discusses the "state-of-the-art" (SOTA) technology, and how the proposed technology of system would either extend or enhance it. This comparison statement should discuss the current limits of the technology, and associated or emulating technologies. Contrast the proposed effort and its potential results with the SOTA. This section also should summarize the current technological capability, point out the weaknesses, and based on the arguments found in the rationale and approach subsections, present a justifiable case for the proposed research. A sample technical rationale is shown below:

Technical Rationale

This section discusses the technical rationale for the Overseer environment, and is organized in three parts. Section 1 presents a technical discussion of the goals and claims for the Overseer environment. The Overseer process and SCVS node structures are outlined. The major functional components of the Overseer environment, including the hardware testbed and load balancing issues, are discussed. Section 2 presents a summary detailing the specific activities and approach needed to complete the Overseer environment and SCVS testbed. Section 3 presents a comparative discussion of the existing SOTA graphical rendering and visualization environments, and the SCVS/Overseer system is contrasted with the proposed research.

1. Technical Justification

1.1 Overseer Basics

The Overseer environment, a rendering and visualization suite for graphically generated and imaged data resulting from numerical computations or sensing devices, is designed to execute on tessellated display subsystems attached to MIMD-class multicomputers. These scalable concurrent visualization systems (SCVS) are, in all effect, multicomputers with each node possessing a frame-buffer and display (such as a small CRT or a LCD flat-panel) device.

The tessellated display elements are entirely isolated from each other, and are therefore incapable of directly sharing data from a common memory store or bus. The proposed architecture for an SCVS is shown in Figure 2.1.

The SCVS has the advantage of being entirely contentionless and totally scalable, thus offering solution speedup and efficiency consistent with MIMD-class multicomputers.

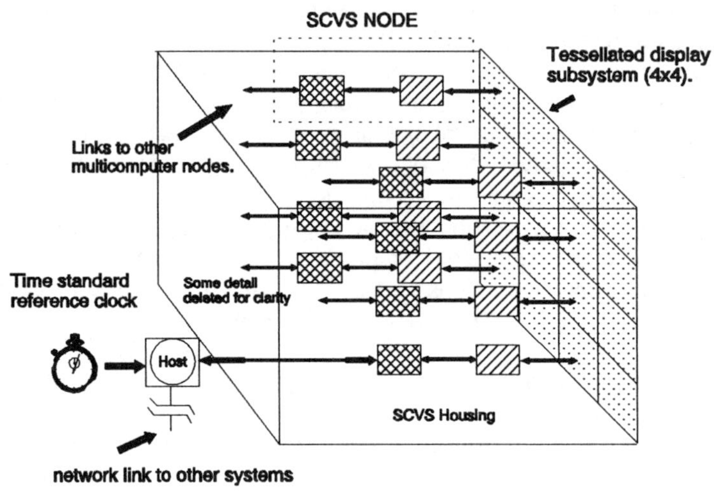

Figure 2.1 The SCVS Architecture.

The SCVS is built from N instances of replicated hardware. The output from the SCVS transformation units (TU) are directed into separate frame-buffers connected to small CRTs or LCDs. Each SCVS TU need only compute a very small portion of the visible scene (the angle subtended within by the viewing frustum on each display element). The perspective viewing frustum is characterized by the field of view angle, the left and right clipping distances, and the near and far clipping distances. Each element of the tessellated display acts as an independent physical viewport subtending $1/N$ of the cumulative perspective view, and this fraction of the perspective view is mapped into each display element as a viewport.

Each display element is thus responsible for rendering (at most) $1/N$ of the total scene, assuming that each TU is load balanced, and the entire SCVS is used. For a tessellated display subsystem arranged as a K by M array of SCVS instances, the geometry will be transformed and rendered as a single scene. The most optimal SCVS utilization will result if all N display elements are simultaneously active, rendering geometry and transforming a scene in a fully load balanced configuration. This condition can be approximated but not always strictly assured in the tessellated environment.

1.1.1 Overseer Process Structure

In a logically concurrent context, the Overseer relies on six processes: the executive, which performs dispatching of the Overseer process structure; the LAMT synchronization process; the SCE load balance process; the GUI/window interface process; the graphical render process, and the viewing frustum/database intersection process.

The LAMT process is discussed in section 1.4, the SCE load balancing algorithm is discussed in section 1.2, the Overseer GUI/window interface process is discussed in section 1.3. The frustum/database intersection is discussed in section 3. Figure 2.2 illustrates the logically concurrent process structure with message-passing interfaces of the Overseer software system.

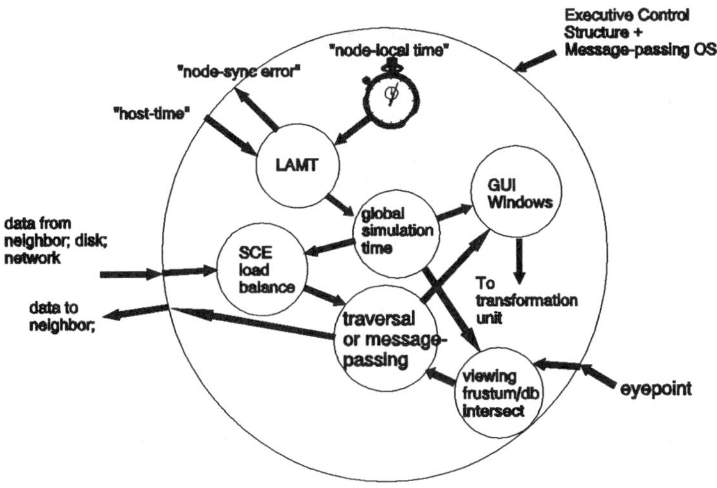

Figure 2.2 The Overseer process structure.

1.1.2 SCVS Node Structure

The endless supply and variety of commodity microprocessors, which will serve as the hub of an SCVS node, provides a vast selection and choice. For building a prototype, the hardware chosen to comprise the three component stages of the SCVS node (traversal unit, graphics process/transformation unit (optional), and display unit) will be selected based on the availability of software support (compilers, operating systems, disk interfaces, etc.). Figure 2.3 illustrates one possible SCVS node structure. The schematic does not substantially differ from that found in an ordinary personal computer or workstation, the message-passing interfaces are the exception. The hardware aspect of the SCVS is trivial in comparison to the software process structure which must be constructed to temporally drive the system.

A multitude of microprocessors candidates are available for the cpu. The message-passing subsystem may be built with off-the-shelf parts, or purchased with the microprocessor. The transputer supplies a cost-effective microprocessor and message-passing subsystem, and may well be used as the principle computation element of the SCVS node.

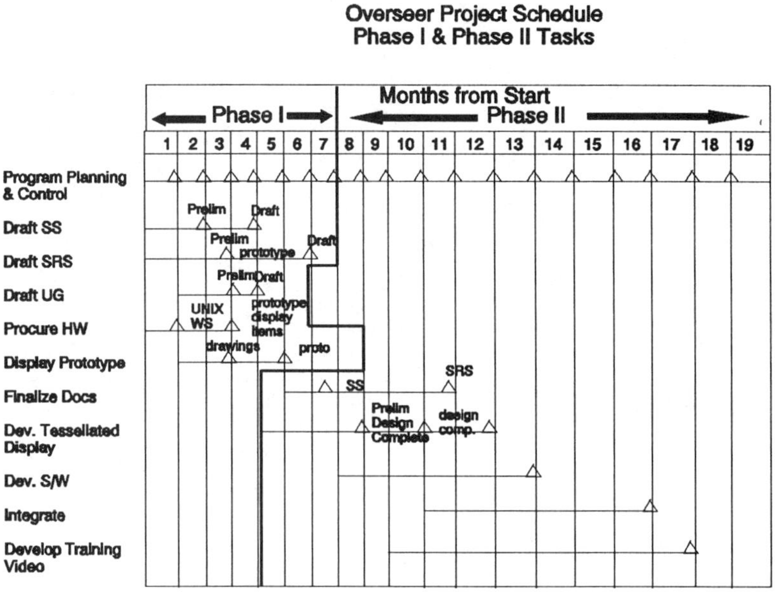

Overseer Project Schedule
Phase I & Phase II Tasks

Months from Start
Phase I
Phase II
1 2 3 4 5 6 7 8 9 10 11 12 13 14 15 16 17 18 19
Program Planning & Control
Draft SS
Draft SRS
Draft UG
Procure HW
Display Prototype
Finalize Docs
Dev. Tessellated Display
Dev. S/W
Integrate
Develop Training Video
Prelim
Draft
Prelim
prototype
Draft
Prelim
Draft
UNIX WS
prototype display items
drawings
proto
SS
SRS
Prelim Design Complete
design comp.

Figure 2.3 SCVS node structure.

1.2 The Load Balance Issue

Figure 2.4 illustrates, for one particular view, how a significant amount of the SCVS bandwidth may, under relatively probable conditions, not be used at all. The instantaneous position of the cup geometry in the world coordinate system, when transformed to the tessellated display, misses several of the SCVS display elements, since the perspective transformation and current eyepoint renders the cup outside the boundary of some SCVS instances. This situation, where the geometry to render is instantaneously precluded from the image plane on specific tessellation instances, is avoided altogether with existing shared-memory visualization systems that possess a standard CRT and frame-buffer, but cannot be avoided with the SCVS architecture.

The tessellated display compounds the load balance issue in one very real sense: a portion of the geometry, while excluded from display in one tessellation instance owing to instantaneous orientation, must be processed by an adjacent transformation unit (TU). This condition, momentary visibility in one tessellation instance and invisibility in another (due to instantaneous eyepoint translation or scene transformation), dynamically alters the local load balance B_1 on each TU, and changes the timing of the update signals for each tessellation instance.

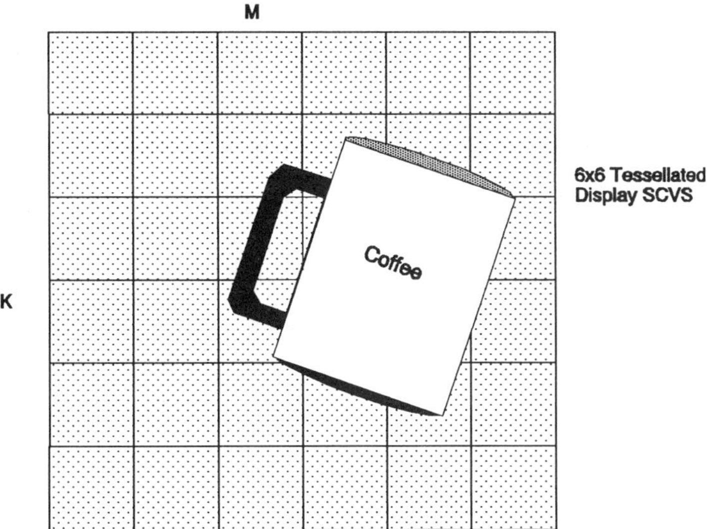

Figure 2.4 The load balance anomaly for tessellated displays.

B_1 is defined as follows:

$$B_l(t) = G_p(t) - G_s(t) + G_r(t) \qquad 2.1$$

where the subscripts p, r, and s denote primitives which are currently node resident, those primitives to receive, and those to send to neighboring nodes. Note that B_1 has units in terms of bytes of storage. Each quantity is a function of time.

The imbalance thus results from one node (or many) passing geometric primitives (polygons) to another for transformation, clipping, shading, and rendering. The time to recover from processing the additional geometry, τ_r, must be less than the global update time $\tau_u/2$ to maintain a real-time transformation rate. It should be noted that if all geometry were replicated entirely on each SCVS instance, no message-passing would be required, but significant visualization bandwidth would be sacrificed. τ_r is defined below, and is directly proportional the amount of geometry in-bound to any node of the SCVS. The greater the amount, the bigger τ_r becomes. This definition implies that node is most heavily loaded determines the frame update rate based on the recovery time.

$$\tau_r = \mathit{MAX}\ \{\tau_{r_i},\ \forall\ i \in \mathit{SCVS\ nodes}\ N\ \},\ \mathit{where}\ \tau_{r_i} \equiv \mathit{total\ message\text{-}passing\ flux\ for\ node}\ N_i \equiv \frac{(G_{r_i} + G_{s_i})}{\beta},\ \mathit{and} \frac{1}{(\tau_r + \tau_s)} = \mathit{update\ frequency}; \qquad 2.2$$

β *is the message-passing link rate (bytes/sec).*

To counteract the instantaneous load imbalance state, and forestall or ameliorate the generation of unstable loads, a constant or functionally identifiable quantity of geometry should undergo message-passing to adjacent nodes. If, for each instant of time, a known quantity of information requires routing from point A to point B within the SCVS (owing to eyepoint translation or a similar transformation), then it follows that the time to perform this operation is precisely known.

This requires that a volume of geometry, if of irregular symmetry and/or structure, or must incorporate fictitious geometric primitives to "regularize" the structure into a highly symmetric scene, or alternatively, that Overseer possess the functionality to detect the distribution of B_1 among all instances and determine an update frequency based on this outcome, which is in turn broadcast to all instances and becomes the new, instantaneous τ_u, τ_u'.

1.2.1 Systolic Compression Expansion Load Balancing

Under conditions when a point-like view of a scene is required (e.g., view from afar), either a different version of the scene must be rendered, one with less fidelity, or SCE load balancing must be invoked. This involves the transmission of geometry and graphical primitives towards a few nodes of the SCVS. A collective in-rush of data forces memory on the target SCVS instances to rapidly fill, forcing τ_r to surge, and this will eventually produce a cut-off in message-passing traffic for primitives. Overseer distinguishes application packets from update signals and operating system (OS) messages, and will continue to transmit these maintenance messages, but cease all others.

A possible alternative to the suspension of message-passing traffic for data would be the introduction of a fast mass-storage device to temporarily store the influx. The mass-storage device, possessing 5 to 10 times the storage capacity of physical memory per TU, could be used to page the in-rushing primitives from perimeter instances until the compression ceases, or the mass-storage device is full.

Following termination of compression, the primitives can be removed from mass-storage in a regular paging pattern, or first-come first-served basis, and then undergo transformation, clipping, shading, and rendering. Mass-storage reclamation can be initiated upon expansion to a normal view, or via a user-defined signal.

Incorporating mass-storage devices is likely to provide a very rapid vehicle for the visualization of large datasets born of multicomputer computation. Placing a disk unit within reach of each four or eight SCVS instances can drastically speed access to large datasets. One of the major shortfalls with existing supercomputer systems is the I/O bottleneck, where several hours may be required to load a multi-gigabyte dataset.

A dataset already residing in perhaps 256 mass-storage units among a 1024 node SCVS, all simultaneously accessible by the connecting nodes, results in a tremendous bandwidth for I/O. Assuming a 5 Mbyte/s small computer systems interface (SCSI) link between 256 of 1024 SCVS instances implies a 1.2 Gbyte/s sustainable I/O rate to mass-storage. Few, if any, shared-memory systems can reach this level of performance. (However, the reliability issue arising from the utilization of this many disk units will have to be carefully assessed and evaluated.) Incorporating a redundant array of inexpensive disks (RAID) configuration is a possible solution.

In the expansion mode, a zoom-like closeup of the dataset or a portion of it results in the replication of only a few primitives on all the instances. Closeup viewing produces magnification on a small portion of the dataset. The expansion mode sustains continuous, uninterrupted message-passing without noticeable degradation in display fidelity because the overall workload node is not drastically increased.

1.3 Overseer User Environment

The Overseer environment will permit multiple users simultaneous access to the SCVS, and this is accomplished through a user interface based on a windowing system. The Overseer environment will achieve this feature by enabling a user to reserve a fraction, N_f, of the SCVS into an arbitrary P by Q subarray. The Overseer window manager will support simultaneous personal visualization in conjunction with other active visualization processes not requiring the entire tessellated display extent. The Overseer user interface will permit standard window functions (open, close, resize, move, pop/push, etc.). These operations can be exploited to support multiple users of the SCVS resource in a networked environment (the LAM environment).

Reservation of SCVS instances into user-definable P by Q subarrays is a feature supported by several commercially available multicomputer operating systems such as the Cosmic Environment/Reactive Kernel from Caltech, or Genesys available from Transtech. Both of these OSs permit a user to reserve a specific quantity of multicomputer computation resources, and the SCVS will have an identical capacity. The Overseer extends this reservation process to include a mouse-driven window interface for activation of specific resources.

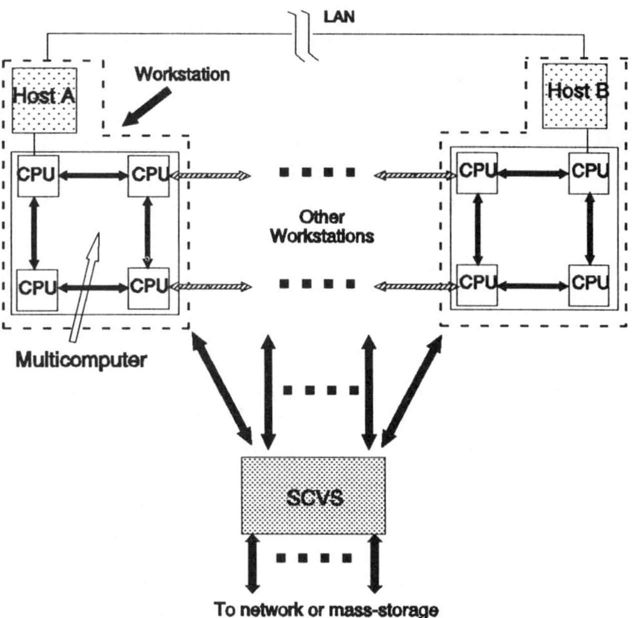

Figure 2.5 The LAM with SCVS resource.

The window interface, and the resulting multi-user capability it affords, will provide an expeditious platform for integrating the SCVS into LAM environment. The LAM is a mechanism for connecting multiple networked workstations into a multicomputer resource (Figure 2.5). In doing so, many users can gain access to a portion of a multicomputer through a remote networked interface. This approach greatly improves a computing facility's resource utilization. The multicomputer resource may possess hundreds or thousands of multicomputer nodes, and not all of them may be simultaneously required by a single application. The LAM makes the entire resource available for simultaneous access.

1.5 Real-time Multicomputers

The concept of time is extraordinarily important to real-time serial computation. It is even more pertinent to real-time computation for multicomputers. Multicomputers are collections of individual computation nodes, each driven with a physically separate clock. MIMD-class multicomputers are inherently asynchronous platforms -- their very design dictates this. But real-time computation demands that time become a universal quantity, and this creates a unique problem: How to create total (bulk) temporal synchrony within a MIMD-class multicomputer?

Realistically, temporally synchronous execution is guaranteed by a SIMD-class computation system. The commercially available machines achieve this goal by an instruction arbiter which broadcasts the instructions to all computation elements. The broadcast structure and network is precisely known and organized to account for all propagation delays while the signals traverse from the arbiter to the nodes. This regular broadcast pattern is maintained by firmware and dedicated routing hardware. To reach a temporally synchronous context in a MIMD platform, an equivalent broadcast pattern must be periodically executed. This activity will establish local area multicomputer time (LAMT) on each computation element.

LAMT is a quantity maintained by all computation elements in the multicomputer, and is recognized as the standard time value from which all temporal events and measurements are logged. A software mechanism for developing LAMT has been implemented on transputer networks. The synchronization mechanism is easily capable of maintaining a global time standard within small transputer networks (up to 16 nodes have been demonstrated), with an error of less than 100 μs by updating the global time at intervals of 1 to 10 s.

It is entirely possible that an algorithm like the one discussed will suffice for networks with more than 16 nodes. Indeed, the Concurrent Computing Consortium's Touchstone Delta multicomputer will have 528 Intel i860 processors, and it would be a great advance for real-time multicomputer systems engineering to demonstrate the behavior of a real-time synchronization algorithm based on LAMT in this system.

Temporally synchronous execution is essential for the Overseer environment. The rate at which each tessellated display element is updated and the temporal phase of the update process are crucial to the generation of a coherent picture. Human factors will not tolerate an update phase delay Δ of more than 33 ms with respect to LAMT, or a 33 Hz phase delay. Δ is determined from MAX {abs (τ_u - τ_i), for all SCVS instances }. This implies that each display element must be updated within ± 33 ms of the LAMT.

The SCVS, therefore, will operate as a loosely synchronous MIMD system, where asynchrony is tolerated provided it does not globally exceed 33 ms. One possible method to enforce complete synchronization would be for all instances to know what their peers are executing in terms of polygon transformations and message-passing. A bitonic sort in a hypercube can be used to rapidly inform all nodes about the load balance and temporal update value. This can be achieved provided that at simulation time t_0, all nodes are equally loaded. At some later times t_i, if each node has knowledge of the quantity of message-passing and local load balance B_l, a single signal may be broadcast triggering a display update.

Any greater discrepancy in the update phase will materialize as a noticeable degradation in the visual scene quality. A visual aberration -- a kind of twinkling effect -- will appear, and could even lead to nausea or motion sickness if the distortion is obvious enough on a large screen several meters across. Predictable LAMT maintenance is a pivotal component of the Overseer environment, and will prevent the twinkling condition.

Embedding LAMT maintenance into the Overseer environment will ensure that accurate time is kept on each node. The signal to update each tessellated display element is driven by the local load balance on each SCVS instance. The local load balance, B_l, of each SCVS instance is determined by the quantity of data residing within the viewing frustum of each display element viewport and the message-passing volume. These factors are linked to the application rendering mode.

In the case of SCVS, the dynamic load balance depends on the rendering mode: static scene with fixed eyepoint (ray tracing), static scene with moving eyepoint (flight simulator), or dynamic scene with moving eyepoint (robotics simulation). The latter two modes require message-passing, and will require SCE load balancing to generate a high fidelity, real-time scene.

2. Technical Approach

This section discusses the engineering activities required to document, design, construct, integrate, and test the Overseer visualization environment as specified in Section 2.1.5 (Statement of Work), and produce the deliverable items stated in Section 2.1.3.

2.1 Systems Engineering

The Overseer visualization environment is an advanced software rendering system which incorporates multicomputer software systems engineering concepts not widely known or popularly practiced (at least in the U.S.). The entire design for a scalable software system, and the SCVS specifically, shall be articulated via logical concurrency. This principle is a method for abstracting the process structure of the Overseer into inputs and outputs (message-passing interfaces), and processes (procedures or functions).

The systems engineering activity will generate a system's requirement document containing the results of a functional decomposition of the Overseer software and hardware components. The decomposition of the Overseer functionality is a first step toward partitioning the system into logically concurrent processes which may require message-passing interfaces. The logical concurrency expresses a measure of parallelism, the degree of simultaneous processing which the Overseer software structure should contain.

The Z specification language shall be used to articulate formal behavioral aspects of the Overseer windowing and rendering software, as well as describe the interactions between the SCVS hardware components at the subsystem level (cpu, framebuffer, graphics processor, external I/O interface, and timekeeping).

The system requirements document will serve as a basis for the computer software requirements document (CSRD). The CSRD contains a specification of the process internals, message-passing interface protocol, and data structures required for the Overseer system. The CSRD will articulate any off-the-shelf products deemed suitable for integration into Overseer. The CSRD lays the groundwork for the computer software design document (CSDD). Also, the computer hardware requirements document (CHRD) will derive functional requirements and purpose from the CSRD.

The CSDD contains the detailed internal design, data dictionary, timing structure, and testbed software interface definition for the Overseer components. The CSDD also specifies the Overseer library interface, a collection of routines which may be integrated into an application program for execution on the SCVS. The CSDD will be used to prepare the Overseer User's guide and training material.

The computer hardware design document (CHDD) will express the schematic and layout of the Overseer testbed hardware. The testbed is needed to verify the execution of the Overseer visualization environment. The CHDD will document the host computer interface, the packaging criterion to sustain large tessellated displays, the mass-storage subsystem connection, and LAM interfaces.

The user's guide and training material will detail techniques for incorporating Overseer library calls into applications, Overseer window manager functions, SCE load balancing techniques, and LAM operation.

All of this material is necessary to construct and adequately document Overseer. The design data, especially for the Overseer software, is especially critical to the effort. The software is the core of this project.

The SCE load balancing concept previously identified is analogous to virtual memory paging to a swapfile on a mass storage unit. One potential mechanism to implement this feature is to simulate the process of paging as a result of in-rushing data streams from remote processes. Instantaneous geometry flow rates may be assessed to determine how much data should be placed in mass-storage, or should τ_u merely be reduced to slow the update process and the entire geometry flow? These decisions are best made following a trade-analysis assessment via simulation or analytical modeling.

The Overseer simulation proposed for Phase I is the second most critical item for the success of the SCVS/Overseer project. As previously stated, the Overseer process structure shall be designed via the principles of logical concurrency (see Figure 2.2). The message-passing interfaces between major processes will be added, and the entire ensemble simulated in a single address space of the procured UNIX workstation using the stand-alone router subsystem of a commercially available OS.

The stand-alone router provides a multi-process emulation of the physical concurrency of the Overseer when it is booted and active on *N* nodes in the SCVS multicomputer. It is far easier to debug message-passing for a particular process structure in a single-address space, then by attempting to "shotgun and debug" a distributed-memory application through individual node address-space examination.

The algorithm and/or techniques used to organize and maintain LAMT are viewed as the centerpiece of the Overseer run-time system. Our transputer network will be used to initiate LAMT algorithm development and testing prior to the hosting of this mechanism on the 3 by 3 tessellated display/SCVS prototype.

3.0 Comparison

This section compares the Overseer visualization environment with existing technology.

3.1 Technical Discussion

Overseer is comparable to any commercially available window interface and GUI. The difference is that Overseer is designed for scalable systems with tessellated displays, and specifically targeted to execute within the LAM resource environment. The LAM support, while already available in some multicomputer OS products, would be the first to target visualization in addition to scientific or numerical products. Providing these two features, scalable visualization with numerical computation, will fill the current void in massively parallel visualization systems. Linear scalable visualization is the next logical step in multicomputer systems engineering evolution.

Commercially available parallel computing systems all depend on a confluence point for conducting visualization. These parallel systems rely on a single frame-buffer into which all computation elements must output results for display. Why these manufacturers have not elected to pursue tessellated displays, like the SCVS, and maintain their extraordinary computation and communication bandwidths for direct visualization is unknown. Perhaps they are frightened of the synchronization problem? Perhaps real-time multicomputer systems equipped with tessellated displays for visualization cannot be engineered? We clearly believe otherwise.

SCVS technology hinges on the tessellated display subsystem, and the generation of LAMT. The Video Wall, an array of monitors showing an enlarged view of a picture imaged from CCD focal plane equipped television cameras, is a close cousin of the tessellated display subsystem envisioned for SCVS. A small prototype is required to demonstrate Overseer and SCVS feasibility for this effort.

But the Video Wall is the converse of SCVS: blocks of pixels are shuffled out of an imaging device's CCD array, through a submultiplexing and pixel magnification system, and then displayed on separate monitors. The SCVS computes each pixel directly, in parallel on separate transformation units, and does not rely on a common store of information. Furthermore, the Video Wall is not scalable beyond a 5 by 5 or 6 by 6 array, since the contention between the CCD interface and pixel magnification subsystems precludes extension.

The tessellated display subsystem of the SCVS is also like the "Jumbotron" giant screen television system built by the Sony Corporation [Bar91]. This system is constructed from a tessellated arrangement of trinilite displays, which contain 256 pixels, and are approximately 1 ft^2 in area. A screen of dimension 40 feet wide and 120 tall can be constructed with this technology, but each trinilite is not driven via an independent processing system. It uses a matrix addressable control system to display compact disk/video or live television signals.

The LCD active-matrix technology is crucial to the tessellated display. Currently the Japanese manufacture high-quality LCD components, but none of the available products accommodates refresh rates beyond 10 Hz. The Coloray Corporation (San Jose, CA) proposes an active-matrix technology which will deliver NTSC-compatible scan rates deemed necessary for the SCVS to be effective (see [Sti90] and [Cor91a]).

We perceive that active-matrix technology of sufficient fidelity will be available within two to three years, the proposed timeframe for Overseer/SCVS development. It is also our contention that display sizes of 256 by 256 or 512 by 512 are sufficient for tessellated display elements. These sizes, consisting of 65536 and 262144 elements, can be manufactured without errors or defects. Constructing larger displays, those with 1 million pixels (or larger) are likely to suffer from a high defect rate.

3.2 Preliminary Performance Estimates for an SCVS Session

A remote sensing and observation platform transmits an image to be analyzed. This image is constructed into a polygonal database of 1 million triangular polygons, for use in a flight simulator/mission training exercise. The remote sensing platform has a field of view equal to 120°, and it has imaged an area equal to 120° by 120° (Figure 2.6). Assuming that all 1 million of these polygons are viewable at each instant, what is the time required for Overseer to render this information under a continuous transformation of rotation through an angle α?

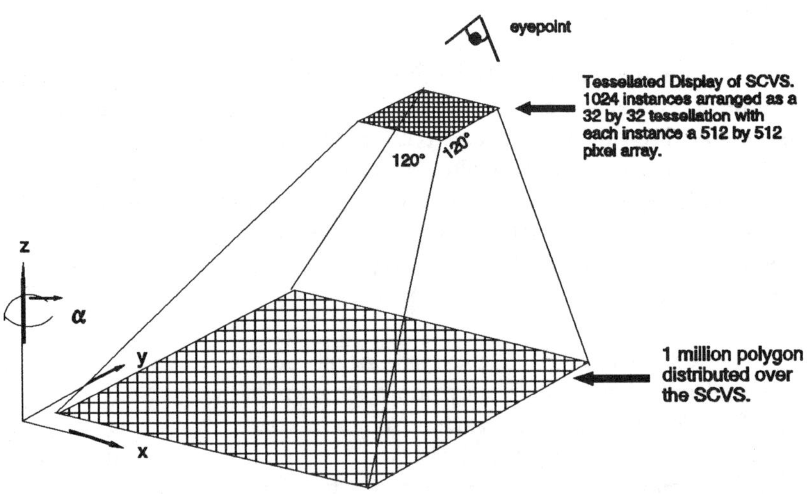

Figure 2.6 Sample geometry for performance estimates.

We calculate this value via two methods: one is a detailed analysis of the process structure required to implement the computation and graphics subsystem functions. This approach subjects the SCVS processing to decomposition. The second method is via an aggregate arithmetic operation count for the graphics subsystem. The method is used to yield various scaling factors associated with CMOS process technology, and provides clues to the potential speedup for SCVS technology as CMOS processes improve with time.

Detailed Analysis

The time to render a scene in an SCVS, τ_r, is the sum of the following quantities: τ_{dbi} -- database and viewing frustum intersection time, τ_{fb} -- the frame-buffer update/access time, τ_{ep} --the time to broadcast the eyepoint/translation data and synchronize, τ_{mp} -- the time to perform message-passing of data between neighboring nodes, τ_{tr} -- the time to transform, scale, rotate, clip, and Gouraud shade all polygons provided that the database is structured as a tessellation (like a flight simulation database). Discussion of these time elements follows.

τ_{dbi} is the database and viewing frustum intersection time. Previous experience with a Silicon Graphics 3000 series Iris has shown that this operation requires less than 10 ms to carry out. τ_{dbi} will likely be less than 2 ms on either the Intel i860 or SGS-Thomson/Inmos T9000 transputer. Since τ_{dbi} occurs on all instances reserved by a visualization process within an Overseer window, the frustum, which is either an orthographic or a perspective projection, must be decomposed into "frustumettes." Each SCVS instance projects a fraction of the entire frustum into the database which is resident on its node. This is accomplished with plane geometry and linear algebra; the algorithms are well-known, and easily derived.

τ_{fb} is the frame-buffer update time. This is the time to shift a block of pixels from frame-buffer memory into the display controller. Each processor has 1024 polygons imaged over a 262144 pixel display (512 by 512). Assume that a pixel is composed of R,G,B and *z* depth information. For each *z*-value read from the *z*-buffer, three frame-buffer cycles are needed. Only one addition is needed for *z*-buffer compare, so three frame-buffer cycles and 1 addition are needed for each pixel. All pixels are visible so 3 * 262144 frame-buffer cycles are needed (786K), and 262144 additions (0.26 Mflop). This requires about 3 ms to accomplish. If LCD technology is used, the finite response time of LCD active-matrix for a pixel to achieve full-intensity is about 50 ms (currently). This constrains the rate at which frames can be generated.

τ_{ep} is the time to broadcast the current eyepoint information (*x,y,z*, and roll, pitch, yaw) to all nodes. This message is composed of 24 bytes (assuming all single precision floating point formats). Assuming a 32 byte message header, 10 Mbytes/s for the message passing rate, and a broadcast to all nodes in the SCVS through virtual circuits created with worm-hole routing chips like CalTech's Mosaic or the Inmos C104, requires 6 ms to conduct.

τ_{tr} is the time to transform, shade, clip, and display 1 million polygons. To calculate this number, the following information is needed regarding the per polygon fill requirements. Each display instance is an array of 512 by 512 pixels. The 1024 instances are organized as a 32 by 32 tessellation. Therefore, each display instance subtends $(2/3\pi * 2/3\pi)/1024 =$ 0.00428 sr. With a coverage of 1 million polygons (each with 3 vertices) in the viewable area, each of the 1024 instances will contain 1000 polygons, and this implies that each polygon subtends .00428 sr/1000= 4.28×10^{-6} sr/polygon. The number of pixels/polygon to shade is given by (262144 pixels / instance)/(.00428 sr/instance) * 4.28×10^{-6} sr/polygon= 261 pixels/polygon.

Table I Arithmetic operation count for graphics function on a per/vertex (*x,y,z,n*) basis.

Graphics Function/Vertex (x,y,z,n)	Arithmetic Operation Total
Modeling from object coordinates to world coordinates	25 multiplications & 18 additions
Trivial view frustum accept/reject	18 multiplications & 14 additions
Simple diffuse and ambient lighting	8 multiplications & 6 additions
Viewing transformation	8 multiplications & 6 additions
Clipping:none (depends on number of non-trivial accepts/rejects)	0
Perspective divide	3 div., 2 mults, 2 adds

If Gouraud shading is considered, then .008 μs/pixel * 261 pixels/polygon= 2.1 μs/polygon are needed for a total of 2.1 ms for 1024 polygons/processor. Each instance will require approximately 32.2 + 2.1= 34.3 ms for τ_{tr}. The term for 32.2 ms is derived by consulting Table I, and applying the operation count to each polygon and assume 100 ns operation time. This is the time required for transformation, scaling, rotating, clipping, perspective divide, etc.

τ_{mp} is the time to transmit the geometry between nearest neighbors. This expresses the amount of time required to move geometric primitives, such as polygons, between SCVS instances, as the scene undergoes continuous transformation through an angle α. The transformation rate is assumed, for this example, to be $\pi/4$ rads/s. Assume that each polygon requires 40 bytes to represent, and that the rotation rate amounts to the physical displacement of 250 polygons per SCVS instance.

This implies that at least 40000 bytes/instance must undergo bidirectional message-passing from one node to another, 25% of the area in each tessellation instance. At 10 Mbyte/s, this implies that 4 ms total message passing time are needed.

T_R is therefore estimated to be: 2 ms + 3 ms + 6 ms + 34.3 ms + 4 ms= 49.3 ms, or about 20.3 frames/s (neglecting context switch and message-passing setup times which are a few μs per/process). This estimate is 20 times faster than the current performance of an advanced graphics workstation rated at one million polygons/s. This estimate neglects graphics processor/pipeline components, which would add considerable speed to the rendering process.

Scaling Calculations

The following operation counts with respect to standard 3D graphics transformations, clipping, shading, and rasterization processes are shown in Table I.

This gives a total of 64 multiplies/divides and 46 add/subtracts per vertex. Figure 2.7 shows the Mflop/processor requirement versus total polygon count (assuming an equal number on each node) required for the 1024 nodes to refresh the display at a 10 Hz update rate. The rate of growth in CMOS silicon process technology is also shown in Figure 2.7. It indicates that a 10 Hz update can be maintained with 1e5 polygons/processor using a .1μ CMOS silicon process. This implies each processor can sustain 300 MFLOPS performance.

Figure 2.7 indicates that by the year 2000, about 1e5 polygons/processor is likely, and in a 1024 node SCVS implies that 1.02e8 polygon databases will be supported at a 10 Hz rate (or 1.02e9 polygons at 1 Hz). The SCVS will thus possess an aggregate processing bandwidth of 0.3 Tflops (scalar).

The aggregate time to transform 1 million polygons (with 1991 silicon) is therefore estimated at: (3 million vertices times 110 flops/vertex * 0.1μs/flops)/1024 processors= 32.2 ms/processor (a 31.0 Hz interactive rate, neglecting message-passing and synchronization which is 15 ms giving about 20 Hz as before).

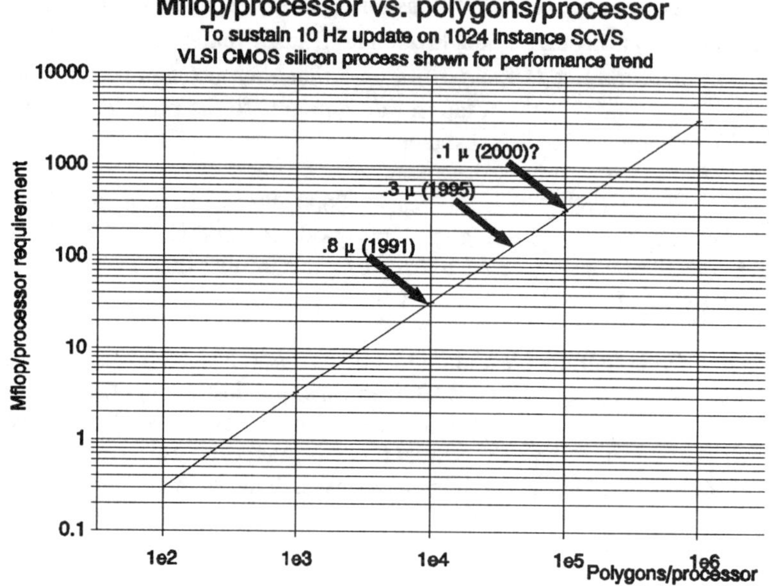

Figure 2.7 Processor speed requirements vs. polygon count (with CMOS process technology trend).

Furthermore, since the SCVS is based on scalable system components, the cost to construct an SCVS is likely to be 10 to 100 times less than shared-memory systems, giving a considerable cost/performance advantage.

2.1.8 Schedule

The *project schedule* illustrates the time-dependency of the milestones and objectives to reach during project execution. The time-dependency is essential, since all commercial or public sector projects require a budget to pay for development. The success a development team encounters during task execution depends on many factors. Skill, knowledge, dedication, management prowess, etc. all play a role when working toward a common goal. The schedule itself must not articulate the impossible.

Requesting one software engineer to design, code, and unit test 10Klines (even with CASE tools, and a reusable component library) in one month is totally impractical. Clearly, the schedule must portend a realizable timetable for execution. The timetable is closely linked with technical rationale, the SOW, the innovative claims, and the deliverable items. A sample schedule is shown Figure 2.8.

The abbreviations for each item are as follows: Draft SS -- write a draft system requirements specification; Draft SRS -- write a draft software requirements specification; Draft UG -- write a draft user's guide; Procure HW -- identify and select hardware items for prototype; Display Prototype -- build the display prototype for Phase I; Final Docs -- after expermimentation with the prototype, revised system requirements specification and software requirements specification to reflect changes and lessons-learned; Dev. Tessellated Display -- build 8 by 8 SCVS; Dev. SW -- build complete Overseer visualization control software; Integrate -- perform system integration on the 8 by 8 and Overseer software; Develop Training Video -- organize training material.

Note, for example, that a milestone marker (triangle) is placed at each month in the planning task. This corresponds to the anticipated delivery of monthly activity reports and progress summaries that explain the work done to date, problems encountered, work-arounds, and other status information. The milestone markers remain unfilled until the actual objective is reached.

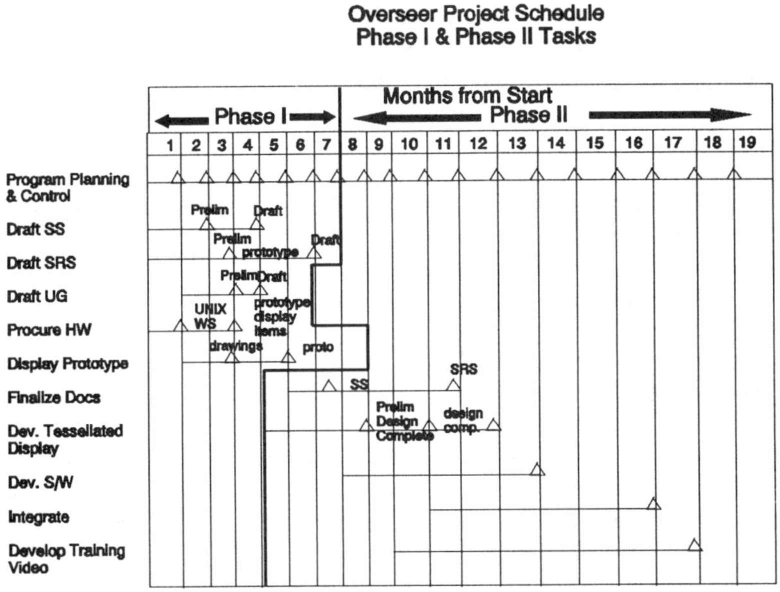

Overseer Project Schedule
Phase I & Phase II Tasks

Figure 2.8 Sample SCVS project schedule.

Also note that the schedule indicates some Phase II tasks start prior to the completion of Phase I. This occurrence is also stated in the cost/schedule summary (section 2.1.3). Some projects require or develop dependencies of this nature, where different phases may be conducted independently (asynchronously) of the earlier phases.

The lines drawn horizontally indicate the duration of execution for any task within the schedule. They terminate with a milestone marker (triangle) on a proposed (estimated) date which is tendered with the proposal. When a milestone is achieved, the triangle is filled in.

2.1.9 Principal Investigator Summary

The *principal investigator summary* lists the vita of the major participants involved in the engineering effort. Each summary vitae should state the name, education, pertinent experience listed chronologically (most recent first), publications, and other exemplary features of the investigator. This summary is used as collateral support by referees to assert that the investigating team possess competency and experience in similar environments and efforts.

2.1.10 Facility Summary

The *facility summary* states the pertinent features of the location where the investigation will take place. The list may include statements like "cleared and authorized for the storage of classified documents," "a 1000 square foot environmentally controlled TEMPEST chamber," "among the available test equipment is a General Dynamics ATE station," etc. The facility summary may include a photograph of the building, or a statement of the square footage.

Concluding Remarks

This chapter presented an overview of project plan preparation methods. The content and organization of a technical proposal reflects the substance of the idea to be developed. The concise statement and enumeration of objectives, cost, and rationale are key to effective proposal writing. Recognition that a project is too complex or expensive to carry out within a complete lifecycle necessitates the incorporation of rapid prototyping methodology to identify and monitor risk factors.

Technologically risky engineering efforts -- those which "push the envelope" of development -- are ideal candidates for rapid prototype methodology. Projects aspiring to develop new SOTA capabilities are difficult to fully assess and specify, especially for an individual working alone. Peer review of the project plan and technical proposal should be considered to improve the technical merit.

Remember, engineering is a collective process. Depending on peers to review and comment on ideas will make them superior, wiser, and more valuable. Strong minds discuss ideas, average minds discuss events, and weak minds discuss people.

The architecture posited by the SCVS should be contrasted with the most advanced, but traditional graphics system research (see Molnar [Mol91]).

Suggested Reading

On the subject of creativity and writing, the reader is referred to the book by Max F. Perutz [Peru89], and that of Raymond Bradbury [Brad89]. The book by Mills and Walter [Mil78] is a good starting point for technical writing.

3

Multicomputer Software Metrics

This chapter introduces the subject of software engineering metrics as applied to multicomputer software engineering projects. Software metrics define the cost requirements and project execution success of an engineering effort. Cost and schedule factors cannot be understated as essential measurements of project performance.

3.1 Metrics and the Software Engineering Process

Correct estimation of the cost and schedule elements of a software project are vital to the successful conduct of the effort. Aside from gauging cost and schedule for a proposal, software metrics are also applied during a project for status assessment. Software metrics are accounting and management tools for discovering problems during the development cycle.

A successfully conducted software effort is characterized by substantially more important goals than purely functional ones. Justification of the claim, "We successfully executed the acceptance test procedure for the first time" is not always synonymous with a successful effort. This claim is ludicrous if the acceptance test procedure occurred two years late and many times over budget. Aside from relishing the fact that an over-budget and behind schedule project is "finally over," a project which is tardy and exceeds cost projections is hardly claimed a success by anyone. Budgetary and scheduling issues weave a unifying thread through an engineering effort. Accomplishing stated or contractual technical goals within the bounds of time and budget is the true mark of a successful engineering process.

Many software projects often fail to meet initial contractual requirements for delivery schedule and budget. Among the most notable in the United States is the modernization program for the air traffic control system, which is over three years behind schedule and US $10 billion over budget [Per91]. While manifold factors contribute to cost overruns and schedule slippage, the key is to prevent these situations from arising altogether. Preventative management methods, such as constantly monitoring key project tasks, go a long way toward minimizing cost and schedule perturbations. An effective combination for avoiding this catastrophe is created by coupling status information with strong technical problem solving skills.

In the same fashion that a physician makes an early detection of cancer, which can be treated and excised from a body, software project management skills serve an analogous purpose. Above all, the problem solving skills necessary to prevent wholesale cost overruns and scheduling problems reside with the initial formulated estimate of the project's budgetary and time performance parameters. Essentially the problem is this: a software effort is needed to complement an advanced hardware system, such as a flight simulator or avionics control system. How much time and money is needed to develop the software?

3.2 Sequential Software Engineering Metrics

Many fine text books on software cost estimation have been written. These texts offer guidelines, procedures, formulae, and methodologies for analyzing a software system to determine, or at least bound, project development costs. Development cost is the most visible component motivating the decision to fund and proceed, or avoid development altogether. Estimating cost for software development is a critical task. More importantly however, is that once a cost estimate has been made, and commitment to execute the project is established, the embarrassment of running over-budget must be avoided. Overruns are a promising source of consternation and headache. To preclude this from happening, studious caution must be applied to the estimation process.

Sequential software metrics have received considerable attention in the United States. Agencies like the Department of Defense (DoD) and the Department of Energy (DoE) are principal purchasers of custom software, and are increasingly dependent on software to satisfy their requirements. The latest estimate for software purchased by the United States Government is enormous, rounding out at US $10 billion annually, and rising. A portion of the US $10B amount is due to mismanaged software efforts. Failure to enforce a structured and positive engineering environment can lead to discontent within the ranks, forcing schedule slippage and poor program performance.

Software engineering is a most human activity. With the skills akin to an artisan, software engineers must have the proper environment to constructively exercise their creativity. They must maintain allegiance to approved methodologies.

Sometimes, the customer changes requirements too late in the cycle forcing a redesign, or the design activity was poorly executed, or the software/hardware technology was not thoroughly understood, or a combination of the above contribute to overbudget software engineering efforts. In cases where the customer drives requirements, they should know better than to modify requirements in the middle of a development cycle.

Often however, the customer does not even know how to specify requirements, and can only offer vague guidelines. In this situation, a large portion of the development effort is used to educate the customer about the product scope, limitations, and constraints, notions which they should be fully aware of prior to extending a request for proposal. Thus, firm requirements definition is the key toward bounding the constraints of a solid software cost estimate. Where requirements are not defined, exercise informed judgement about creating them, but above all define them!

3.2.1 Software Lifecycle

The software lifecycle is an engineering process defining the various stages of software development and maintenance phases encountered during the lifetime of a software application. A lifecycle is characterized by distinct stages, each one fulfilling a specific role in the lifetime of the software application.

The most popular and successful lifecycle model formulated to date is the "spiral" model [Boe87]. This model stipulates a precise relationship between a system's functional objectives, and the methods applied toward the realization of an operational product, through the subsequent enhancement of the system, to the system's retirement from operational service.

In particular, the spiral model holds that once a system's feasibility has been established, and the system specification is complete, several prototypes should be constructed through an iterative development cycle, until the system is made operational. The spiral model mirrors realistic software engineering practices, as conducted on a daily basis throughout US industry, especially by contractors in the aerospace industry.

3.2.2 The Spiral Model and Rapid Prototyping

The spiral model points out a particularly important notion for development efforts which must evolve along with changing technology. Since the model calls out a specific pattern of prototype engineering prior to the synthesis of an operational system, it implies that iterative design and refinement is employed to continually improve the functional objectives of a prototype.

Eventually the prototype is regarded as operational after several iterations. And after each stage, more functional objectives are progressively satisfied. Following delivery of the operational system, enhancements to the original delivered system are applied. This activity may include the wholesale porting of the original system to a faster platform, or merely a tuning activity designed to exploit specific advantages of an operating system.

During either the prototype or enhancement stage, opportunity exists to refine and modify existing capability for newer or more functional characteristics. Clearly, these modifications can only be realized if the system is organized via a configuration management process, so that controlled upgrades can take place without irreparably destroying the initial functionality.

For systems with long-term serviceability, such as the data system used to analyze interplanetary streams from a satellite system with constantly changing observation and measurement volumes, it is clearly more sensible to design the system to evolve along with the data stream requirements. Thus, rapid prototype development of a data reduction system is a logical method to use in this instance.

The rapid prototype process can be facilitated at smaller institutions and firms that are not burdened by excessive overhead and bureaucracy. In some ways, the current procurement practices of large funding agencies do not permit rapid prototyping, as this practice runs counter to the established engineering methodology of fixing requirements, freezing design, and building to a specific design. While technology evolves, contractual stipulations hinder an equally innovative approach to creating topical solutions.

Regulatory bottlenecks result in the construction and introduction of technologically obsolete products long after the initial design is frozen, and development/manufacturing takes place. This is especially true for many items constructed for military purposes, where many years pass before an operational system can be introduced. By the time the system is readied for deployment, a new generation of hardware is available which can easily overpower the old. While standards and technology rapidly evolve, regulations change slowly and create unnecessary roadblocks to progress.

3.2.3 Spiral Models in Neophyte Organizations

The author has witnessed more than one large software system irreparably damaged by eager engineers, who acted outside the scope of a configuration management process, and "added" one new feature to the system which could not be backed out. This perturbation led to an expensive system-wide upgrade of the entire software application. The cowboy-like attitude resulted in a modest expenditure of US $400,000 in additional labor to correct.

This quantity of money is small by the standards of a large corporation, but the fact that a failure in the management structure allowed this to occur at all points out how critical effective configuration management tools are to the development process.

The spiral model is an important contribution to software engineering practices, but to affect development within this structure requires a fully developed software management control environment and system. Otherwise, the spiral model can invite disaster for the uninitiated organization.

3.3 COCOMO

Barry W. Boehm [Boe81] has authored the definitive text on sequential software metrics. The principle theme of his text describes the COnstructive COst MOdel, or COCOMO. This model has been derived from the measured performance of several small (less than 10Klines of code), medium (10 to 100Klines of code), and large (greater than 100Klines) scale software efforts. It is not a panacea, but has distinguished itself as the foremost representative technique employed by the software development community within the U.S. The author has participated in many structured environments where COCOMO provided the quantitative cost estimation figures used to bid projects.

COCOMO has two variants known as Basic and Intermediate. Basic COCOMO estimates a development effort based on the size of a project only. It produces worker-month (WM) effort (the equivalent of 152 working hours) as a function of deliverable source instructions (DSI).

DSI excludes nondelivered support software, such as a special in-house code used for constructing global databases, and also excludes documentation efforts for users or internal design data not officially covered by the product development cycle.

Intermediate COCOMO estimates cost according to the principle cost drivers of a development effort (complexity, reliability, software engineering skill, computer hardware factors, execution requirements, etc).

Intermediate COCOMO parametrically derives cost estimates through consideration of tangible software engineering environmental factors. Basic COCOMO is a simple equation. Both are useful to gauge and quantify project costs.

This chapter is geared toward Basic COCOMO, and is intended to only introduce the topic of software metrics. For a more rigorous treatment of the subject, consult [Boe81].

3.3.1 Basic COCOMO

This form of COCOMO, termed Organic, is simply stated as:

$$WM = 2.4\ (KDSI)^{1.05} \qquad \textbf{3.1}$$

where WM is worker-months,[15] and KDSI is thousands of deliverable source instructions. The quantity KDSI includes documentation beyond any internal code comments. This equation expresses the development metric, in worker-months, for a project of known size. Provided that your code estimate is true, this equation gives an estimate "within 1.3 of the actual project execution only 29% of the time, and within a factor of 2 only 60% of the time [Boe81, pg. 84]." Therefore, one should regard the estimates generated from Basic COCOMO with cautious skepticism.

Equation 1 has three variants: the Organic, Semidetached, and Embedded. Each variant of Equation 1 uses slightly altered coefficient and exponent values. The variant which is applicable to a project depends on several factors, but the application itself is key to the metric employed for estimation and assessment purposes.

The predictive accuracy of Basic COCOMO is derived from the COCOMO database, a collection of over 60 projects, each scrutinized, measured, and monitored for performance. Basic COCOMO estimates and metric expressions are probabilistic in their predictive success, but they affirm and represent typical software engineering efforts.

This internal historical performance database is tailored to a specific environment with quantified attributes, strengths, and weaknesses, and may be used to greater success for Intermediate COCOMO. A seasoned organization generally enjoys a greater probability of success than the novice one, since managerial and engineering personnel evolve a bond of trust and coordination -- teamwork -- which is often unsynthesized in the neophyte.

Whether an organization elects to incorporate either Basic or Intermediate COCOMO, the benefit of investing human resources into the discovery and development of customized metric estimation parameters rather than failing to recognize the importance of cost estimation by miserably botching an engineering effort far outweighs the expense.

---

[15] Worker-month is a gender-independent label for engineering human resource.

3.3.2 Work Breakdown Structure

Basic COCOMO covers tasks specified by a work breakdown structure (WBS). The WBS details specific structural items of the software product to some level of detail. A WBS is extremely useful for partitioning a product development effort into manageable elements which are identified with levels of effort, activities, and cost. It is far easier to estimate project cost once the entire structure and extent are known, and the WBS is a helpful mechanism to quantify resource requirements.

An example WBS is shown in Figure 3.1. For this project, the scalable concurrent visualization system, several computer program components and supporting engineering activities are required and outlined in the various components of the WBS.

The WBS in Figure 3.1 is a very modest example. The WBS for the space shuttle, the B1-B bomber, or even an automobile is exponentially more complicated and articulated. The larger the project -- in terms of assemblages, components, connections, interfaces, functions, and requirements -- the larger the WBS naturally grows. Entire careers are made on estimation and the organization of WBS.

Each element within the WBS may be assessed independently of the others. Partitioning a project into smaller, more manageable phases and elements aids estimation. Expending the effort to organize a WBS, and refine estimates based on this organization, is extremely useful.

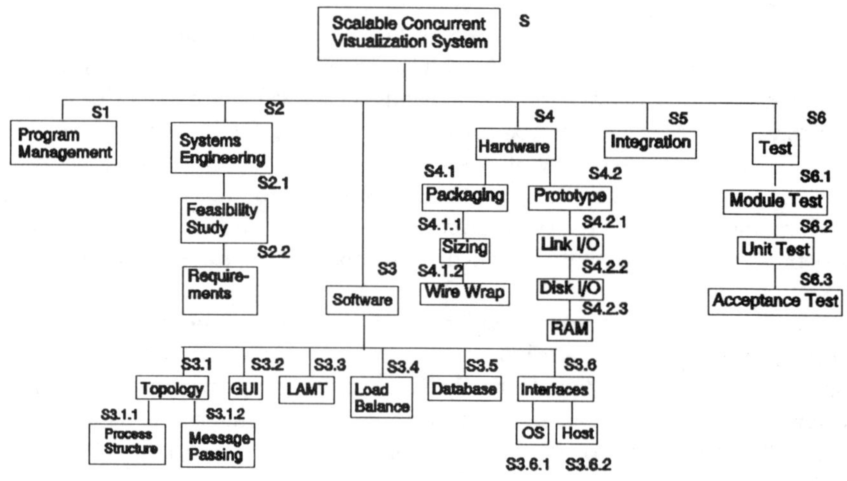

Figure 3.1 Work breakdown structure.

Firstly, breaking out specific tasks gives visibility to the entire effort. One can more easily assess and estimate a task once it is clearly defined and the scope is articulated. Secondly, a WBS is an essential component of most planning activities.

Identification of the WBS will help to prevent misapplication of resources.

With Basic COCOMO, an accurate assessment of the program source code requirements are essential. Like all estimation models, or equations in general, uncertainty in the input produces uncertainty on output. Determining the number of worker-months effort to deliver a specified quantity of anticipated code or product identifies the labor cost for the principal effort.

Support for configuration management, quality assurance, code maintenance, librarians, and other costs are directly anticipated with Basic COCOMO. But indirect overhead expenses (janitor, computer costs, marketing personnel, secretaries, etc.) are not accounted for by Basic COCOMO.

Overhead expenses are an important factor which distinguish a small business from a large one. Fewer personnel are found in smaller firms, and the overhead is correspondingly lower. This can lead to a competitive advantage for bidding on and winning contracts.

The human resource requirements are determined by multiplying the number of worker-months for a project by their monthly cost. This value determines the total dollar value for the labor needed to perform the software engineering for a particular project. Add to this the computing costs (if the machine is leased), the hardware cost (if a special process is being built to support the project), equipment leases, etc. All of these factors -- and many others not mentioned here -- are often included into the project cost estimate.

3.3.3 Calendar Estimation

How much time, in terms of calendar months, will be required to complete a software project? Basic COCOMO provides an equation to estimate the development schedule in calendar months. The equation:

$$TDEV = 2.5(Q)^{0.38} \qquad 3.2$$

expresses the total calendar months of development required to carry out a lifecycle on a software project estimated to require Q worker-months of labor. Equation 1 and Equation 2 are a tandem of tools. They should be used together to gain an approximate bound on the engineering effort for a particular project. Clearly, the accuracy of these equations in the production of an estimate is governed solely by the precision of the system size assessment.

The total calendar time estimated by Equation 2 can aid with the decision to fund an effort. If time is a critical issue for a particular software product, owing to competition in the marketplace, this equation provides information which may promote a project, or cancel it. Equations 1 and 2 can provide valuable baseline information for a project requiring X calendar months to complete, and Y worker-months of human resource. This baseline data can be used to construct an expression for the workerloading effort, and the timing of human resource application to the project.

Workerloading estimates are extremely useful to determine staff requirements. One must be certain that a fixed number of individuals are needed to support a project. A workerloading estimate determines the human resource based on the projected requirements.

3.3.4 Workerloading

Derivation of the worker-month effort requirement for a project is closely associated with the worker load estimate, the allocation of personnel resources during the execution of the project schedule. Workerloading defines the quantity of human resource to be applied during the various project phases. In a structured software environment, such as the one spelled out by DoD-Std-2167 [DoD87], several phases are identified as project milestones. The project milestones are goals that signify completion of work according to a schedule.

Figure 3.2 illustrates a sample workerloading schedule for the scalable concurrent visualization system. The workerloading schedule illustrates how, in a temporal sense, the manpower for a project will be applied to the engineering process. Note that during the requirements phase of the project, an estimated 2½ persons are needed. This implies that two full-time engineers are needed for the duration of the requirements definition process, and an additional engineer is needed part-time to support this activity.

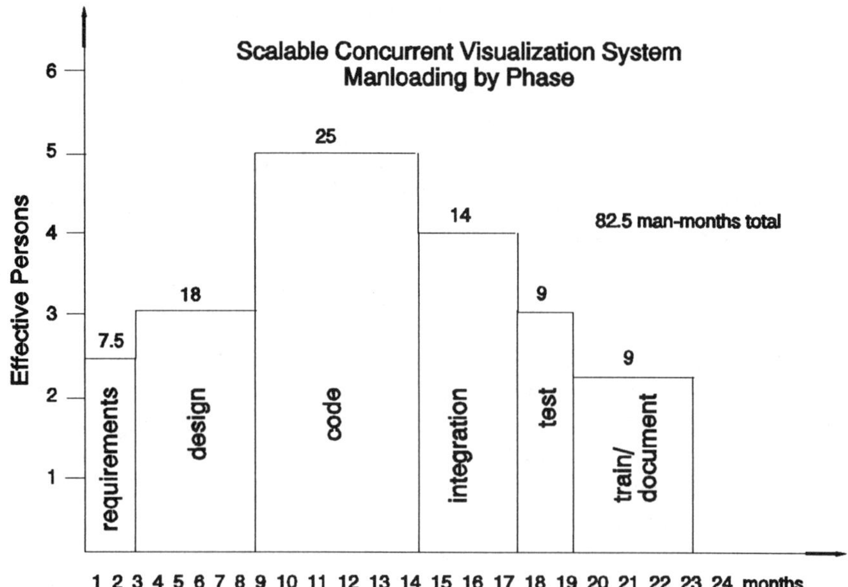

Figure 3.2 Workerloading schedule.

To compute the total estimated worker-month effort required to execute the software engineering process from beginning to end (requirements definition through training/documentation), one merely integrates the area underneath this curve. The total labor cost of engineering personnel can be found by multiplying the area (worker-months) by the company-derived worker-month cost factor. Thus, a typical aerospace firm may determine that, on average, an engineer's cost to the company on a worker-month basis is US $10,000. From Figure 3.2, this gives an estimated labor cost for engineering personnel at US $825,000.

A more realistic cost estimate is given as a weighted sum of the worker-month cost, since not all engineers cost exactly the same for each worker-month of service, a reflection of salary disparities and other tangible factors. Typically, a system engineer's time may run at twice the rate of a software engineer, while a program manager's time is 3½ times that of a software engineer's. Many government contracts require a labor breakdown statement by the hourly or monthly cost of project personnel.

Another attribute of the workerloading schedule is the ratio of time resources allocated for requirements definition versus those needed to design solutions for the requirements. A rule of thumb for this ratio is 1:2½ or 1:3. For each hour of requirements definition, approximately 2½ or 3 hours are needed to design a solution to a requirement.

Obviously, if the software engineering environment is well established and robust, meaning that a large repository of existing software modules are available as documented and tested entities ready for inclusion to construct other systems, less resource will be needed for design, code, integration, testing, and documentation.

A critical fact to remember about workerloading estimation is that if an area of a project falls behind schedule, extreme caution should be exercised about staffing up a particular task for the recovery process. As the saying goes, "one carpenter can build a dog house in one month, but thirty carpenters can not build one dog house in a single day." This adage applies to software projects as well. When a task is languishing, take the time to assess the causes of lessened projected performance, diagnose the problem, and act accordingly. Perhaps extra manpower is not the answer, but a better CASE tool might resolve the issue.

3.4 Multicomputer Development Cost Assessment

Multicomputer software systems are a unique breed, and are quite distinct from sequential or shared-memory multiprocessor software systems. A multicomputer software system is engineered, in today's world at least, when no other off-the-shelf computing system can satisfy cost goals and performance requirements. The reason for this is simple: a data parallel application executes N times faster than the same sequential solution, and typically costs 1/10th to 1/100 that of the equivalent shared-memory platform on a Mips/Mflops basis.

While platform costs are still of tremendous concern, the arrival of the microprocessor ("killer micros" [Mar90]) has propelled computing platforms to greater performance levels for less expenditure. A typical microprocessor-based workstation can be purchased for a few thousands of dollars (less than US $10,000) which possesses significant computing power. But a workstation's computing power is only harnessed by the software which directs the solution. Software has become the critical commodity satisfying the demand of consuming industries.

The cost assessment strategy presented in this section is based on software engineering economics derived from a basic comparison of multicomputer execution contexts. The two primary contexts are data parallel and control parallel simulations.

3.4.1 Data Parallel Software

Data parallel software is typically the simplest and most cost-effective variant of multicomputer software. It is simple in the sense that the identical calculation or process is executed or applied to each portion of a decomposed dataset, which is assumed to be distributed throughout the multicomputer. This arrangement can substantially reduce the software engineering complexity. Since the same calculation is performed, this means that only a single subroutine or computational algorithm must be specified, designed, coded, tested, integrated, and documented. Only the message-passing interface is needed to complete the multicomputer implementation (see Figure 3.3).

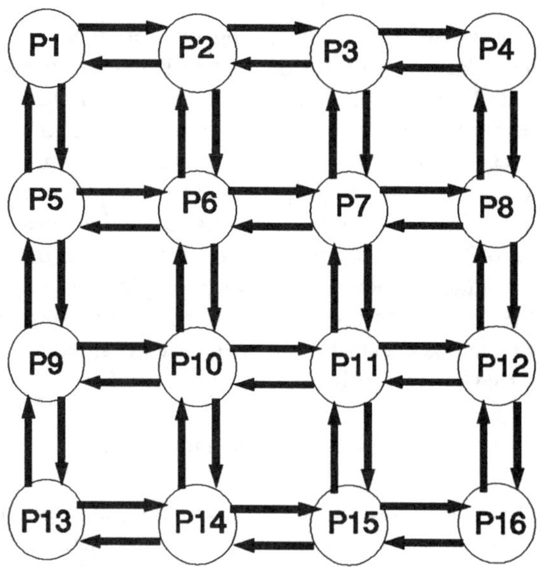

Figure 3.3 A data parallel process structure.

Data parallel simulations are scalable. They are constructed from a single-program multiple-data (SPMD) approach: the replication of a single process throughout a multicomputer platform which operates on a decomposed data set. This form of software architecture is most efficient for MIMD-class multicomputers, and gives rise to maximum speedup under the right circumstances (see Chapters 6 and 7). A scalable process structure is entirely reusable. Provided the problem domain is large enough, it is entirely possible (and indeed quite achievable) to replicate a data parallel process structure ad infinitum, provided enough processor resources are available [Gus86].

From the standpoint of software costs, a data parallel simulation is the cheapest possible to implement. Fewer calculations imply less software, and less software to write requires fewer people to design and write it! The message-passing interface is a simple edifice to construct. It is literally a few lines of software which conveys data from one node in the multicomputer to another.

The ease with which a data parallel simulation can be organized and ported to a multicomputer depends on the intrinsic geometrical symmetry (or lack thereof) in the dataset.

For example, if one writes an algorithm to compute the propagation of thermal energy through a solid, the multicomputer implementation is eased in the case of highly symmetric problem geometries, such as grids, parallelepipeds, spheres, and cylinders. These coordinate systems have well known symmetries, and they are easier to decompose into a regular multicomputer topology which matches the problem geometry.

Irregular domains, such the structural model of a ocean-going passenger ship, the airframe of a jet aircraft, or the architecture of a medieval cathedral, are much harder to coerce into a multicomputer structure. These irregular, non-symmetric geometries create load balance anomalies which, although data parallel, require non-intuitive techniques to balance correctly (see Chapter 7 on Load Balancing).

In these cases, the message-passing edifice assumes a less obvious structure: rather than simple nearest neighbor communications between processors, data might be routed from distant nodes, and this implies that a data tracking and identification scheme be used to locate a datum's position.

3.4.2 Control Parallel Software

The other variant of multicomputer application software are the control parallel simulations. These simulations are characterized by M processes, each of a unique purpose and function, and are highly dependent on message-passing interfaces. They are limited in the degree of assumable parallel execution due to internal dependencies or their architectural structure. A flight simulator, or an avionics control system for an air-vehicle, or the environmental control system for a large office building are examples of control parallel simulations and applications.

Each process in the control parallel simulation possesses multiple functional modes that depend on information -- messages -- from different control threads. The relationship between the threads of control precludes a decoupling of the processes. This binding limits the amount of achievable speedup that a decomposition or replication of the structure can attain.

A multicomputer implementation for any of these examples requires that many unique processes be subjected to the software lifecycle. Thus, whereas the data parallel case requires a single process to undergo development, the control parallel case possesses a degeneracy factor of M. The M processes become multipliers applied to the sequential software engineering metric posited by COCOMO.

Each of the M processes is unique in scope, function, memory, timing, and size requirements. To estimate the time and cost required to develop each, Equations 1 and 2 must be used separately in each case, and the results added together to produce a final cost and schedule estimate.

While the degeneracy factor is a notable element of control parallel multicomputer simulations, it should be pointed out that shared-memory simulations obey similar metrics, the principle exception being that in a shared-memory implementation, the effort required to achieve a load balance is substantially lower than in the multicomputer version. However, ease of load balance attainment does not necessarily make the shared-memory system superior, only easier for most software engineers, who are trained in the traditional and classical sequential development scenarios, to construct. Additionally, there are far more engineers who are trained in the classical, sequential software engineering methodology than for the multicomputer realm.

It is the control parallel simulation which offers the greatest software engineering challenge. The challenge arises from the confrontation between obtaining maximum speedup in solution, and creating a confounding design by introducing unnecessary complication. Techniques for extracting maximum speedup from control parallel simulations are discussed later in Chapter 6.

Each process in the control parallel simulation is characterized by at least one message-passing interface, where a precisely known message packet data structure is either sent or received (see Figure 3.4). This is also true of data parallel computations. But each unique process in the control parallel case, should it have a data dependency between more than one peer, will possess an equal number of message-passing interfaces, one for each peer sharing communication.

The communication interfaces constrain the potential speedup achievable from the simulation. It is not so much that message-passing is hugely expensive (in most multicomputer systems, it is actually quite efficient), but that these interfaces express the data dependency of the process structure graph, and each message that is transmitted or received creates a synchronization point for the simulation. The fewer the number of synchronization points, the less time wasted waiting for communicating processes to exchange messages.

This scenario is similar to that found in a shared-memory multiprocessor, where one process is allocated to each processor. With the addition of an extra process to the job queue, a parallel context for each of the original *M* processes will suspend until the operating system permits the new process to execute for a timeslice.

The operating system enforces an execution order, and governs timeslices for all processes of equal priority. This forced synchronization in the control parallel multicomputer simulation is common to all message-passing systems; processes execute asynchronously, and only coordinate (synchronize) when messages are exchanged.

3.4.3 Comparison of Metrics

Since the control parallel simulation possesses a factor of *M* degeneracy over the data parallel case, and the most significant goal of a multicomputer simulation resides in the potential to obtain an *N* times solution speedup, removing the degeneracy from control parallel simulations is a seriously important priority and design goal. This task, unfortunately, is not always an easy one to expedite. Attempting to extract or coerce a data parallel simulation from a control parallel architecture via automated means (e.g., a compiler or other software tool) is an unsolvable problem.

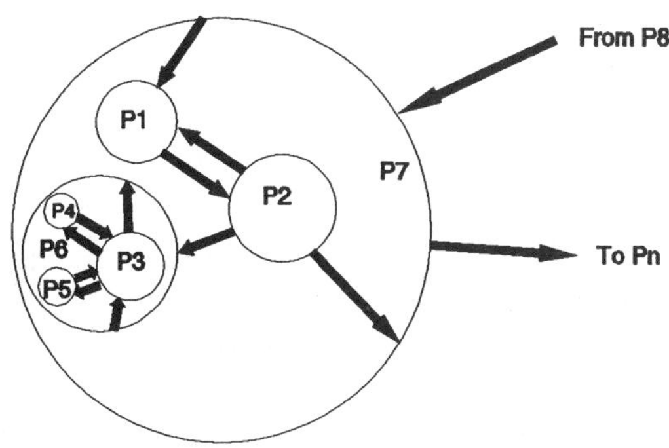

Figure 3.4 A control parallel process structure.

The task of performing automatic parallelization via a compiler has not yet been achieved. This notion is equivalent to the Turing Problem [Bern66], and establishes the likelihood of such a tool as an entirely intractable case. Hence, only manual means will prevail if a control parallel simulation is to realize some of the potential solution speedup so evident from data parallel simulations.

The metrics for data parallel simulation software obeys the rules established by COCOMO models. A single process, executing a single computation or algorithm, equipped with a solitary or highly structured and symmetric message-passing interface is engineered in precisely the same manner as a sequential algorithm. In fact, the algorithm used in a data parallel simulation may be precisely the same tried and true algorithm constructed many years hence, only augmented with message-passing structures.

In the control parallel case, it is usual that a block of software consumes the bulk of the execution resources, and this bottleneck receives the focus of attention as attempts are made to optimize it. The so-called "80-20" rule -- 80% of the computation resource is spent in 20% of the code -- indicates where data parallel simulation software may be used to effectively eliminate this anomaly. Locating the areas of the software where high resource utilization requirements exist, and redesigning the algorithm to affect a more data parallel structure, can alleviate the processing bottleneck.

But what expense must be incurred to adjust the algorithmic structure of a control parallel application into a data parallel context? This question is extraordinarily difficult to answer. Given the state of software engineering tools for multicomputer systems, an immediate solution providing an estimate for the conversion is not likely. The best approach is to begin the design process with data parallelism in mind.

The dilemma arising from the motivation to obtain faster solutions via a data parallel context for existing software, and the difficulty encountered during the porting effort stands as a great barrier preventing many applications from realizing a scalable implementation.

Concluding Remarks

This chapter has introduced the basic notion of software metrics for use as an estimation tool for cost and management control. The COCOMO model developed by Boehm is used extensively throughout the software industry in the U.S. The applicability of this metric to multicomputer software systems is straightforward. In the data parallel case, one employs COCOMO to the lifecycle of a single process. The control parallel case requires *M* applications of COCOMO workerloading and calendar estimation equations, once for each unique process in the application.

Suggested Reading

For an extraordinarily lucid and candid discussion of software engineering practice, the reader is strongly encouraged to examine *The Mythical Man-Month* by Frederick Brooks [Bro75]. Cusumano's book on the software factories of Japan [Cus91] gives an interesting account of this nation's approach to reducing software development costs.

4

Introduction to Real-time Computer Systems

This chapter provides an introductory overview of real-time computer systems. Functional characteristics of sequential real-time computer systems are used as the foundation for discussion. Sequential systems share many similarities with multicomputer systems, especially from the perspective of software. To understand and eventually derive a real-time multicomputer system design, a thorough knowledge of sequential real-time systems is essential.

Most engineers are familiar with computer simulations. Computer-aided engineering programs that predict a structure's response to an applied load, or an algorithm to compute spreadsheets or prepare income tax returns or derive stock market averages are all examples of computer simulations: they simulate the process of a system, once chiefly performed by a human being, through the execution of a computer program. Most computer simulations are not real-time, they are "fast" simulations; a solution is computed as fast as possible by the platform.

More often than not, fast simulations run in the background on a big time-share system, or execute during off-peak hours because they demand substantial machine resources to produce a timely result. Computer simulation has become the hallmark of modern society, since a simulation is often far less expensive to maintain and operate than the equivalent staff of humans performing a similar function.

4.1 Definition of a Real-time System

In general, a computer system has two distinct components: the computation, or algorithmic specification of the computation described by software, and the underlying platform which hosts the computation. A *real-time computation* is defined as an ability to conduct and realize a finite computation over a fixed and periodic time interval. A *real-time computation system* is the platform or electronic mechanism for producing a real-time computation. A simulation executing in real-time provides a known and essentially constant level of *interactivity*: the rate at which inputs, computations, and outputs are serviced, derived, and completed. A real-time system must be *predictable*. A deterministic quantity of computation and input/output must be reliably generated on a continuous time-dependent basis.

Stankovic and Ramamritham [Str90][16] articulate a real-time system as a composite of five characteristics. The first characteristic they define is a "granularity of the deadline and laxity of tasks." The *granularity deadline* describes the urgency of a task's execution and completion. If a task must initiate and complete execution within a time interval which is close to the operating system context switch interval, then the deadline is said to be *tight*, and implies a small granularity deadline. Conversely, a *loose* or large granularity deadline can arise where a task's initiation time -- response -- is tight, but the time required to complete the computation is large in comparison to the response.

Secondly, the "strictness of deadline" attribute describes the *value* of a computation after the deadline for a task has expired. The value of the computation is temporally dependent, and loses significance if the result is late. The depreciation of a real-time value is application-dependent. Clearly, if the application is highly critical, such as an avionics control system, each depreciated value from the computation can contribute to an unstable and dangerous function. Depreciated value is unacceptable for *hard real-time* applications, like the avionics control system; the value is either discarded, or some corrective action is applied to minimize its effect. But a less critical application, such as an automated teller machine, may merely generate inconvenience to the consumer if the deadline expires. In these *soft real-time* applications, the depreciated value is not discarded, but salvaged for some diminished purpose. The strictness of deadline criterion has profound consequence for real-time software systems safety (see Chapter 5).

Thirdly, the "reliability" of a real-time system is characterized by a subset of tasks termed "critical." The critical tasks are designated by their contributions to the overall safety and functional performance of the system. Their deadline strictness and reliability are usually guaranteed to be repeatable. "If a critical task misses a deadline, then a catastrophe may occur [Str90, pg. 248]." Stankovic and Ramamritham argue that many hard real-time systems treat all of their composite tasks as critical, thus creating an over-specified and over-designed implementation -- a costly mistake.

Fourth, the "size of system and degree of coordination" required for real-time systems varies. Most real-time systems are entirely memory-resident, which places all tasks within reach of the processor, and the interaction among tasks is usually fixed and well-defined. Systems which have well-defined phases, if they are too large for memory resources, can be swapped or loaded in anticipation of the execution sequence with a multiprocess structure. However, next generation systems, such as a platoon of Mars rover robots, may function with highly dependent communication and computation structures. The independent and autonomous operation of next generation systems may not be economically feasible owing to the excessive design complexity needed to support fully autonomous function. The coordination and size requirements of next generation systems may complicate a system design.

Finally, the "environment" where the real-time system operates places further constraints on the design. High-stress environments, such as the heat generated from

---

[16] This essay is a terrific piece of analysis and definition of real-time systems. The reader is strongly encouraged to examine it.

engines, or the extreme chill of outer space, are often treated as deterministic by the designer. "These environments give rise to small, static real-time systems where all deadlines can be guaranteed a priori [Str90, pg. 249]." These systems are built via off-line analysis to determine timing properties (e.g., counting source instructions).

But future real-time systems may not operate in deterministic environments. Again consider the platoon of Mars rover robots. They will confront unique terrain and climate conditions which will change with the seasons and geography. Rather than a preprogrammed repertoire of steps to carry out, these systems must adapt to a variety of conditions. Mechanisms for adaptation, such as evolution, must be incorporated into these machines to ensure survival and nominal function. How is predictability defined for a system operating in a non-deterministic environment?

Stankovic and Ramamritham [Str90] have outlined two approaches for realizing predictability in real-time systems. To ensure predictability, both approaches focus on engineering design activities, and confront real-time systems design problems from different perspectives. The "layer-by-layer" approach considers a microscopic investigation of the hardware response details, operating system and support software performance, and other factors that affect task deadlines. By assessing these system features, a bound on the worst-case performance can be discerned. This method is similar to many existing design approaches (including those experienced by the author) to assembling performance estimates for real-time simulations.

The second method is the "top-layer approach." Here, the focus is to consider the predictability requirements for the software application. "Deadlines imposed on activities other than those that occur at the application layer are artifacts. The lower layers need not be predictable, but only provide services so that the application layer is predictable [Str90, pg. 253]." The top-layer approach is used for complex activities, where the decomposition of tasks and functions may not be possible through a layer-by-layer approach. Both methods outlined in [Str90] are worthy of consideration as heuristic guides to building a predictable real-time system.

4.1.1 Classification

Many real-time embedded computer systems have a mixture of hard and soft timing constraints; non-embedded systems can also possess these timing characteristics. Real-time systems with hard timing constraints are designed to generate results which are formally or mathematically correct, but the validity of the computed results hinges on their prompt temporal delivery as well. Thus, a *hard* timing constraint demands that the arithmetic or logical simulation results are mathematically correct, but their arrival must also be predictable.

A *soft* timing constraint arises in asynchronous process contexts. The arithmetic or logical results must be correct, but can be late. Missing or overshooting a timing deadline does not disqualify the result. The fast simulation discussed above can be interpreted as one form of a soft timing constraint, albeit an extreme one. A soft timing constraint for a real-time system is likely to arise from a collection of asynchronous processes that are dispatched to execute at regular rates, and attempt to complete prior to the next scheduling interval, but do not always meet the deadline; they execute a lazy context.

Embedded computer systems are predominantly real-time with hard timing constraints. Every jet aircraft contains dozens of embedded control systems for a multitude of purposes. If your automobile has electronic fuel injection, it is controlled by an embedded real-time computer. An *embedded* computer system performs a specialized and autonomous role as a satellite within a larger system, such as a data acquisition unit on-board an aircraft, and may communicate the results or status of its function to a supervising computer system for evaluation.

The importance of hard timing constraints in a real-time system cannot be over-emphasized. Imagine if the flap controller of an aircraft wing did not possess the response bandwidth to smoothly raise or lower a flap in response to the pilot's command at the control stick, but instead jerked around the commanded position for several seconds prior to settling into a nominal position. This airplane would certainly provide a disquieting ride for the passengers! Hard timing constraints often lie at the core of safety and performance for many large systems that depend heavily on embedded computers to conduct successful missions.

Variable-cycle jet engines require updated monitoring and control signals at a maximum interval of 20 to 50 ms, or they can become unstable and destruct [LHA91]. Wherever dynamic systems require controlling agents, real-time computer systems fulfill this vital role. Before an embedded system becomes a realizable component of a larger edifice, its functionality, performance, reliability, maintainability, safety, and other salient features are often simulated in a laboratory environment as part of the design process.

4.2 Design Environment Practices

The success of a real-time computer system, whether embedded or not, hinges on the ability of the participating engineers to deliver a real-time simulation that executes on a given platform. The platform may be a custom engineering job, such as the units found in many commercial aircraft, or an off-the-shelf computer. In many instances, the platform has little bearing on the software design process. Unless it is a mission-critical embedded system requiring extreme fault tolerance and reliability features, standardized commercial computing platforms are suitable, and modern software engineering technology permits near platform-independence for the simulation design and implementation. Thus, the *physical configuration* of the platform, how the I/O and cpu are arranged or packaged, has little bearing on the software engineering.

However, the platform's primary role is to provide adequate computational resources and interfaces for I/O, which is why these requirements are often satisfied by almost any commercially available system. The *logical configuration* of a host platform, the architecture, describes the I/O and computation structure which must be programmed. Establishing a formal specification for the logical configuration is essential for the software engineering of the simulation (see Table 4.1).

Aside from its role as a computational resource, the host platform is typically called upon to provide the simulation development environment. In the case of an embedded system, a suitable surrogate environment often substitutes in this developmental capacity, with the resulting simulation software cross-compiled into an executable image and uploaded to the target system, or the simulation may be stored into EEPROM, and simply plugged into the target.

The later approach was used on the B-1B bomber (built by Rockwell International) as this method was deemed essential to the successful execution of operational flight programs (OFPs). It is far safer, easier, and less costly to simulate the OFP while the aircraft is still on the ground, than to discover an error in the OFP midway through a critical mission. The space shuttle's OFP is subjected to a similar procedure [Fey88].

Since the simulation is the critical item to develop in most cases, it follows that the software engineers who develop the product must have an adequate set of tools to complete their tasks.

Table 4.1 Configuration attributes of real-time systems.

	Platform Configuration Attribute	
Attribute	Physical	Logical
I/O (serial, parallel, optical, electrical, RF, etc.)	Mirrors interface to device under influence of platform.	Message-passing protocol between device and platform needs structure.
Computation (serial, parallel, fault tolerant, etc.)	Redundancy increases tolerance and raises cost; computation speed bounds fidelity of simulation.	Simulation design should be scalable and platform-independent; fault tolerance burdens software engineering effort.
Packaging (power, size, etc.)	Embedded platforms often have severe packaging constraints on power and volume consumption.	Minimal impact for software; power management for embedded systems is generally uncomplicated.

4.2.1 Real-time Engineering Environments

The development environment holds the key to a successful systems engineering effort, but is especially critical for software-related tasks. No matter how fast the hardware may execute, a poorly designed and organized simulation will embarrass the platform performance by not delivering optimal capability. Creating a near optimal capability during the systems engineering phase for a real-time simulation implies that concerted effort, nominal expense, and extraordinary discipline are invested to achieve an excellent match between the hardware resource and simulation software.

In this case, a near optimal match between the hardware platform and simulation software is taken as that quantity of engineering (funding, scheduling, human resources) needed to derive a finished product which satisfies all functional performance, safety, reliability, and quality objectives while not exceeding the available resources.

Since few organizations have unlimited budget and schedule constraints, a compromise occurs. The system requirements and design are fixed, and development proceeds under the assumption that the engineering constraints are justifiably correct and amount to the smallest possible risk from a financial and technical perspective.

This scenario is often attempted in the United States, especially within aerospace environments. One need only list and compare the development costs and initial program objectives versus the final expenditures for government procured projects to assess the accuracy of "The best laid schemes o' mice an men/Gang aft a-gley" [Bur58].[17] The planning and management elements have been previously discussed in Chapters 2 and 3.

4.2.2 Development Tools in Real-time Systems Design

The tools needed to conduct a real-time engineering effort form a unique set of requirements. The uniqueness arises from the temporal aspect of real-time systems. The engineer participating in a real-time systems engineering effort must ensure mathematically correct computation and simulation, and that the results are temporally valid. The temporal dependence supplies additional constraints on the engineering environment. Extra-thorough efforts on the engineering staff's behalf become the norm, rather than the exception. The results of these efforts materialize during the integration and test phases where the "proof is in the pudding."

The tools used to integrate and test, as well as design or prototype a real-time system, must possess the requisite fidelity to support these activities without introducing distortion or skewing the measurement process.[18] This is a tough requirement to satisfy, even in a classical real-time system, but a clever use of computing resources and other mechanisms can supply a good approximation of the phenomenon under investigation.

4.2.2.1 Hardware Tool Requirements

In the case of hardware development, a good logic analyzer, multi-channel digitizing oscilloscope, network analyzer, breadboard equipment, and electronic computer-aided design (ECAD) software for schematic layout are mandatory. Circuit simulation software is helpful to prove truth-tables and I/O logic prior to breadboard layout; it

---

[17] Translated as "go astray" from Scottish Gaelic.

[18] The Heisenberg Uncertainty Principle states that one cannot absolutely know the entire state of a system or a particle (in the quantum mechanical world), as the measurement of a system causes perturbation. Classical systems have no restriction.

also can help to prevent excessive jumpers from appearing during hardware/software integration. Waveform editors are helpful to simulate temporal logic flow. An in-circuit emulator (ICE) is also handy for embedded systems. ICE platforms provide host/target cpu emulation via a combination of hardware/software simulation routines and structures. They often come with sophisticated tools to debug embedded systems, and are generally accepted as standard practice in most system houses.

Any digital design engineer who knows his or her trade will almost certainly ask for these instruments. If the system involves reference to precision time and voltage standards, then an atomic clock and precision power supply are necessary. There is often no substitute for the equipment deemed essential. Instrumentation or support equipment economies may jeopardize the project through failure to meet performance objectives.

4.2.2.2 Documentation Requirements

For software, however, the lifecycle is principally driven by documentation efforts arising from requirements, design, and testing. The lifeblood of any reputable software effort is measured by the state of the documentation. How does the documentation reflect the clarity and content of the product design? To what extent are error conditions, machine dependencies, and global variable accesses enumerated? The quality of product representation is the main contribution of documentation to a software effort.

The United States Department of Defense has articulated a standard, DoD Standard 2167 on "Defense System Software Development." DoD-Std-2167 outlines a sequence of activities and documents required to support, document, itemize, and account for the software engineering activity. This standard is incredibly rigorous, and contains many documents and activities which are burdensome in a commercial environment.

An alternative to DoD-Std-2167 development practices are the American National Standard Institute (ANSI) and Institute of Electronic and Electrical Engineers (IEEE) software standards. This collection includes plans for software configuration (ANSI/IEEE Std. 828-1983) and quality assurance (ANSI/IEEE Std. 730-1984) management, software test documentation (ANSI/IEEE Std. 829-1983), software engineering terminology (ANSI/IEEE Std. 729-1983), and specification development for software requirements (ANSI/IEEE Std. 830-1984). See [IEE83].

The ANSI/IEEE standards specify a less rigorous and demanding methodology for organizing and executing software lifecycles. In many instances, the methodology is an acceptable alternative to the rigors of DoD-Std-2167. However, DoD-Std-2167 requirements are often tailored to suit a particular engineering environment, in which case the conduct of the practices becomes more bearable, and less costly to implement.

One important issue DoD-Std-2167 intrinsically supplies is a mechanism for developing software metrics. The ANSI/IEEE standards have no provision for this important aspect of the software lifecycle.

4.2.2.3 Data Dictionary Editor

Aside from the documentation effort, the organization of global databases which the real-time computations act on to reproduce the desired system behavior also requires treatment. A *data dictionary editor* (DDE) is a tool which facilitates the construction of a global database.

The DDE, and the database it constructs, often serves as a nucleus of understanding for the participating design team members. An integrated picture of the simulation modelling can be learned from an examination of the data structures and simple variables operated on by the simulation algorithms. In many instances, a software librarian chairs meetings of the design team, as the members discuss and outline the basic structures needed to construct the database and reproduce the desired phenomena at run-time.

The librarian will usually operate the DDE and maintain the database content through the process of database meetings and approval cycle. Each design team member will contribute forms to the librarian for creation, modification, or deletion of entities in the database. The DDE output can usually be directed to generate the core of a database description document, thus satisfying one requirement in the DoD-Std-2167 lifecycle model.

One additional facet of DDE functionality is the construction of subordinate or ancillary tables and dictionaries. A DDE which constructs a *flat symbol table*, a hierarchical data structure that associates the physical byte-wise offset of each globally scoped variable used in the simulation, can be exploited by many utility programs (see section 4.4.4 on Monitoring). The global database is a concatenation of variables, of differing types and sizes, against which the simulation functions and processes are linked.

A flat symbol table is a reflection of this global database structure. The database is loaded into memory at run-time, but a flat symbol table has knowledge only of the offset of each variable from the beginning of the database. Thus, while the relative number of bytes separating the address of each variable from each other is known, the absolute (physical) load point is not known until the database becomes memory-resident.

Hence, the program section address value corresponding to the database head is acquired during initialization to establish the relative offset and address of each variable in the flat symbol table. This simple address computation can be used at the core of a software monitor which implements an interpreter and simple grammar. The interpreter provides the capability to symbolically parse the grammar, which includes variables names as a part of the syntax. The symbolic monitor is discussed in section 4.4.4.

4.2.2.4 Precompiler

Another useful tool for development efforts is the language *precompiler*. A precompiler detects variables which are locally instanced within a compilation unit, and prepares an include specification containing the equivalences and common global references for the compilation unit. The compilation unit is then submitted for compilation and linking into the entire simulation.

The precompiler is generally useful for simulations which are based on languages possessing static bindings and flat data types, such as FORTRAN.[19] In large simulations, where thousands of variables are manipulated by hundreds of functions and subroutines, a precompiler saves the developer substantial time by including and inserting the correct organization of common blocks and equivalences.

The precompiler executes by scanning and comparing a compilation unit's declaration and variable usage with the contents of the database organized by the DDE. In most structured environments, variable definitions and usage will have predetermined naming conventions and nomenclature for variables, so the precompiler can be built from a simple lexical analysis specification [Les78].

4.2.3 Design Practices for Sequential Systems

Designing a real-time computer system from the ground up is one of the most challenging prospects an engineering team can confront. Before the design process can be initiated, the system requirements must have been established. For the hardware element of the system, the digital design engineers who will build the platform will most certainly need a good ECAD program for schematic and wiring list preparation.

If the platform involves many custom programmable elements, such as application-specific integrated circuits (ASICs), or other electrically configurable logic elements, then a circuit simulator program, such as one based on IEEE Standard 1076 [IEE76] may be suitable to simulate the logic functions before "burning the silicon." VHDL simulations are tremendously useful for verifying the functional attributes of custom-designed silicon. Simulations of this kind provide substantial cost savings. Changing a few lines of software to move a few gates around is far less expensive than building a new mask for compiling a device.[20]

VHDL simulations do not necessarily execute in real-time, but they can exercise the logic functions, clocking schema, and other pertinent device elements. The entire VHDL simulation process constitutes a formal algorithmic proof of the circuitry which is then compiled into silicon.

For effective and correct software simulations, formal specification practices and their implementation through specification languages, such as Z, provide an equivalent method for mathematical verification of requirements and function. Constructing software systems depends on the process of abstraction, design, code, integration, and test. In a formal environment, the requirements and design phases of

---

[19] Emerging dialects of the FORTRAN standard, such as FORTRAN-90, permit data structures as part of the language extensions. This feature gives FORTRAN a better image, but it still falls short of the Ada, C or C++ languages for abstraction purposes.

[20] The same can be said for the traffic lines painted on highways. Once the paint is applied, its very difficult to remove it (sandblasting is needed). The analogy with silicon processes is apparent.

a software lifecycle are expressed and conducted in a separate specification language. Document preparation is carried out by constructing the requirements or design specification in the formal method grammar accompanied by an explanatory paragraph of two. In most cases, the formal specification grammar is not suitable for general purpose programming; compilers do not translate the specification into executable code. The primary role of the formal specification is to unambiguously express requirements and functional design.

Formal methods seek to construct a mathematical representation of a system, which is derived from set theory and predicate logic. The resulting specification can be mathematically proved, thus verifying the function and intent of the design. Formal methods are closely allied with the important goals of software safety and reliability. Embedded systems, whose failure or ill-mannered function can result in serious loss of life, property, or physical impairment deserve careful consideration and diligence to specify, design, and build. Formal methods eliminate the uncertainty from standard verbal specifications, like those expressed with the English language, and this advantage over existing practices can prevent errors from creeping into the later stages of the lifecycle.

Formal methods of software design can be implemented in several ways. The most expeditious is to actually simulate the software specification with a special purpose tool [Win90].[21]

Another method for designing software involves the use of computer-aided software engineering tools (CASE). Organizations that can afford a CASE toolset conduct their design activities within the environments established by the CASE vendor. CASE toolsets are generally self-contained; they possess the utilities for maintaining audit trails and configuration management for all elements of the software design process. This includes the construction of databases, truth tables or state transition diagrams for logic flow, and data flow diagrams for all hierarchies of the software system.

CASE tools provide a consistent advantage for software engineering, especially for medium or large projects (20 Klines to 50 Klines for medium; 100 Klines or more for large). Many individuals and teams must cooperate on medium and large projects, and a common framework for organizing documentation, requirements, and design configuration is often essential to the coherent and successful execution of an engineering effort.

Without CASE tools, documentation for the software engineering process becomes *free-format*, and more difficult and expensive to maintain. In a free-format documentation process, the maintenance activities for updating documents, essential products of the software lifecycle, consume greater human resources, and the potential for introducing error becomes larger. CASE tools maintain an automated hierarchical document management scheme for tracking specification changes and modifications, a requirement for DoD-Std-2167 environments. This is a distinct advantage over free-format systems.

---

[21] CADiZ from the University of York is such a tool for the Z specification language.

CASE tools for low-end personal computers generally cost less than the equivalent product hosted on a workstation. Currently, about US $1000 is typical for a widely available CASE toolset with modest functionality [McC89]. The kind of CASE tool an organization elects to employ depends on the project size and the liquidity of funding for infrastructure investment.

4.2.4 Design Practices for Scalable Simulations

A scalable simulation, depending on the quantity of control parallel constraints and process structure, can obviate the rigidity and requirement of CASE tools, as a simplified design process may suit the application. This likelihood implies that free-format documentation practices may be an acceptable alternative. Why expend capital on tools which are not necessary to accomplish the job? Data parallel simulations are more likely to fit the free-format documentation development cycle. However, control parallel simulations contain a substantially articulated process structure, and thus demand stricter software development practices and discipline.

A data parallel simulation is generally simpler to design, for a single process is replicated. Design technique plays a pivotal role in developing a scalable simulation. A thorough treatment of multicomputer software engineering design technique is given in Chapter 6.

4.3 Real-time Simulation Structures

A real-time simulation, in the sequential, uniprocessor environment, is constructed from several distinct components. Aside from the application which generates the time-evolution and progression of the simulation, support functions are needed to create the infrastructure that the simulation uses for timing, I/O, event logging, interrupt handling, and other essential elements. A simulation application will not sustain or promote a *deterministic* behavior, a known level of performance which is reproducible for a fixed set of input conditions, unless the execution environment provides specific, predictable, and reliable run-time features.

Real-time simulations often employ an executive control structure for this purpose. An executive provides task timing and *dispatching*. The dispatcher is the preeminent responsibility of an executive, as this function meters the rate of task execution. A *task*, as defined for this discussion, may consist of a separate context, as in a complete and isolated process composed of separate compilation units: modules, functions, or subroutines containing numerical, logical, I/O and initialization codes, and other processing responsibilities.

A simulation is therefore a superposition of tasks, each designed to generate or reproduce specific operational attributes associated with an existing environment, problem, or phenomenon. Figure 4.1 illustrates the logical structure of a simulation.

The inputs and outputs correspond to external events or commands from other devices or tasks within the simulation. They drive the course of the simulation, just as the pilot's control stick drives the flight controls of an aircraft.[22] A real-time simulation must have at least one clock to maintain a standard global simulation time, the reference against which all events and processing occur during the simulation's lifetime.

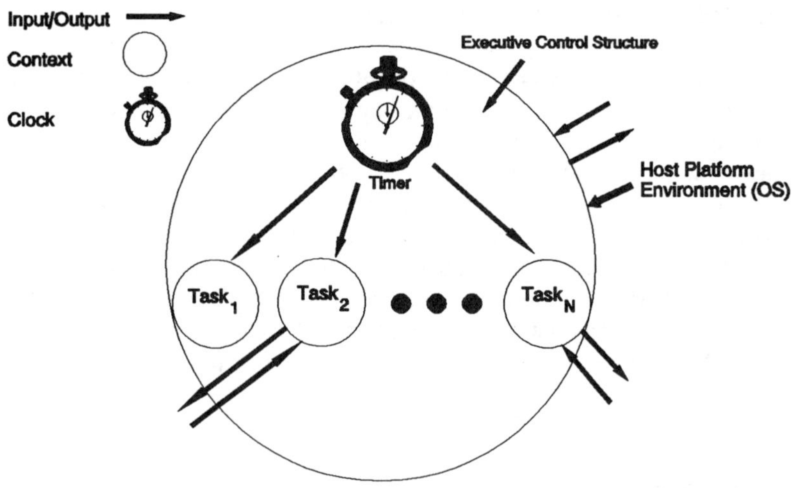

Figure 4.1 A simulation is composed of inputs, outputs, processes, and clock(s).

Figure 4.1 distinctly shows the instantiation of a single clock. The implication of a single clock in this figure is that the entire simulation occurs within a single address space.[23] This depiction of the logical simulation structure infers that a single processor platform is used to run the software. Although there is no reason to suspect that a shared-memory multiprocessor system will ultimately host and execute the simulation.

In a multiprocessor shared-memory configuration with *N* processors, a single

---

[22] This situation is rapidly changing in commercial aircraft transport. Fully automated cockpits with inertial and satellite navigation, and "fly-by-wire" control systems relegate pilots as redundant backup. When the first failure of fly-by-wire occurs, a big backlash in FAA regulations will follow, forcing a return to pilot-in-the-loop flight.

[23] A single address space means that a contiguous block of memory contains all process contexts, data, and registers. Virtual memory in a uniprocessor environment functions by mapping contexts between secondary storage (e.g., disk) into a single address space prior to the resumption of a process context.

clock source is used to coordinate all processing activities. The logical structure shown in Figure 4.1 is equally applicable to shared-memory multiprocessor configurations provided that a single clock source is used to control memory access patterns and process scheduling.

That only one clock is illustrated in Figure 4.1 has bearing on the possibility for implementing this logical simulation structure in a distributed memory multicomputer system. For a multicomputer is configured with *N* processors, but each possesses a physically separate timing source. A mechanism for maintaining the simulation time within a distributed memory computation system is an active area of research. Chapter 8 on synchronization discusses this subject at length.

4.3.1 Executive

The executive lies at the core of a real-time simulation. This element is often dependent on the host platform operating system (should one exist), or at the most extreme is hooked into the host microprocessor through specific traps, register structure, or even the physical addressing modes. The most portable executives are written in high-level languages such as C or Ada. Choosing a portable language to implement a simulation executive will result in terrific savings of engineering resources when a new platform arrives (and they always do!).[24]

The simplest embodiment of an executive is illustrated in Figure 4.2. The executive structure shown here contains an implicit loop (of indefinite iteration) over which the individual tasks are dispatched.

Each task is described by a task descriptor template,[25] an initialized instance of a data structure used to construct the task table. The task table body lists instances of tasks (functions) to invoke.

The executive obtains a periodic interrupt from the platform clock, and this interrupt impinges on a process which contains *N* tasks. Each time the process obtains the interrupt, the task currently pointed to by the task table index is executed, causing computations and/or I/O to be initiated or completed. Thus, each time the executive obtains the interrupt, the process under its control conducts some computation or I/O -- depending on the task table body contents -- for a predetermined period or *quantum* of time (each Δt seconds). The task table body is cycled through by incrementing the task table index.

---

[24] For multiplatform portability, the simulation may be better served by licensing a third-party executive. This approach eliminates maintenance, but raises capital expenditures for the development process. If many copies of the simulation are sold, the capital expenditure is usually justifiable. Almost every third party operating system for real-time purposes supplies POSIX 1003.1 compliant (IEEE Std. 1003.1). The proposed real-time extensions appear in the draft standard for POSIX 1003.4.

[25] Often referred to as a task control block by operating systems engineers and designers (which I am not).

Under non-real-time conditions, Δt corresponds to the time-slice of the processor, a quantity related to the granularity of the system clock. For certain systems, Δt may be 1/60 of a second, while on others it is 10 ms or finer.[26] A process will not be permitted continuous access to the processor for more than Δt seconds, after which time, the OS will conduct a context switch, and a new process may execute.

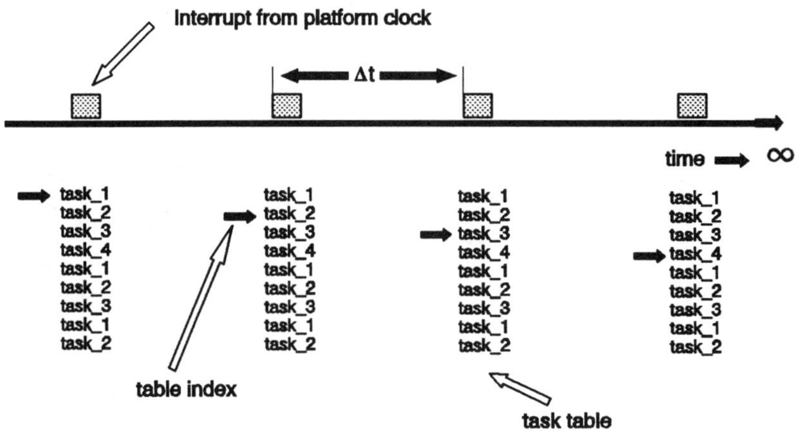

Figure 4.2 Executive control structure with periodic dispatching governed by the platform clock.

A process may be intrinsically I/O-bound, and will typically undergo many context switch cycles prior to completion. This also depends on the host platform I/O support, and whether it possesses an intelligent controller for disk or device I/O.

Figure 4.2 illustrates two important concepts. Firstly, since the process containing the task table is periodically interrupted -- which implies a context switch -- each task must complete before the next interrupt. A *purely synchronous* real-time process depends on a *temporally deterministic* computation, one which is known to correctly complete its algorithm or function within a predetermined interval. Synchronous processes are ideally slaved to the platform clock, and conduct exact quantities of the simulation during each scheduling period; they are statically predictable, as discussed earlier.

Figure 4.3 illustrates the task table structure for a purely synchronous real-

---

[26] The transputer from SGS-Thompson/Inmos gives two timer ticks, at 1 μs and 64 μs.

time process task structure. The task table body data structure descriptor contains atomic entries used to articulate the task context. The task name and the address of the task to dispatch denote a primitive context in this example. If necessary, other atomic attributes may characterize a task table entry, such as a variable for the last scheduled time of execution, the next time period for execution, and an execution mask which may cause a task to be periodically disabled, etc.

```
typedef struct
{
   char task_name[32];
  long (*task_address)();
} Task_descriptor;

Task_descriptor task_entry_table[];

task_entry_table[] = {
                { "task_1", task_1, },      /* multiple instances of identical tasks */
                { "task_2", task_2, },      /* implies a higher invocation context  */
                { "task_3", task_3, },      /* frequency                             */
                { "task_4", task_4, },
                { "task_1", task_1, },
                { "task_2", task_2, },
                { "task_3", task_3, },
                { "task_1", task_1, },
                { "task_2", task_2, },
         };

#define NUMTASKS ( sizeof(task_entry_table)/sizeof(Task_descriptor) )

extern long task_1();
extern long task_2();
extern long task_3();
extern long task_4();
```

Figure 4.3 The task table body and requisite data structures for a purely synchronous, real-time process context.

Secondly, the number of tasks instanced by the task table body governs the task duty cycle. A process which is dispatched by a platform interrupt with a period of Δt seconds coincides with a nominal dispatching rate of $1/\Delta t$ Hz. Within a process, each task instance within the task table body garners processor cycles equating to a fraction $\Delta t * a_i$ seconds, where a_i is the total number of instances for a task within the task table body.

The total frame time for a process, T_f, the quantity of time required for the task table body to cycle once, is given by Equation 1:

$$T_f = \sum_{i=0}^{N-1} a_i \tau_i \quad \text{4.1 Frametime determination.}$$

where τ_i is the amount of time required for each unique task to execute. The frame rate, Ω_i, for an individual task τ_i within a frame requiring T_f seconds to complete is defined by Equation 2:

$$\Omega_i = \frac{1}{\Delta t} * (\frac{a_i * \tau_i}{T_f})$$

4.2 Execution context frequency for task T_i within a frame requiring T_f seconds to complete.

Equation 2 illustrates the functional dependence of the platform clock interrupt period on the task frame rate Ω_i. If task τ_i requires a very high execution fidelity, clearly one must decrease the clock interrupt period and possibly increase the number of instances of τ_i within the task table body.

According to the Sampling Theorem [Bra78], a function cannot be reproduced from sample values obtained at a frequency less than $2f_c$, where $2f_c$ is known as the Nyquist frequency, and f_c is the highest frequency of interest in the sampled function. The Sampling Theorem has very profound consequences for real-time systems, especially those designed for signal processing applications.

Should a task within a real-time application conduct waveform reconstruction, then it must possess fidelity greater than $2f_c$ ($2\Omega_i$), or the reconstruction process will be based on aliased information, and produce misleading and altogether incorrect results. Thus, over-sampling is almost always necessary in the absence of signal conditioning filters, as noise tends to pervade a signal, distorting the fundamental harmonic structure of the desired waveform.

The Sampling Theorem holds broad implications for embedded real-time systems. Recall our example of the flap controller, and how, if the computer system cannot continuously and smoothly sample the pilot's control stick commands, the flap is ill-behaved. This aliasing phenomenon is a manifestation of either an ill-designed simulation or an entirely faulty embedded control system platform.

4.3.2 Multiprocess Executive Structures

The discussion thus far has purposely excluded the notion of multiple, simultaneous[27] process contexts. Multitasking operating systems specialize at permitting many processes to share machine resources. These OSs are the backbone of many timeshare computing installations. They can also serve as adequate hosts for many real-time simulations.

Multitasking OSs can be exploited to yield effective real-time simulations, provided the simulation hosted on the platform does not overwhelm the kernel by demanding too many simultaneous processes. An instantiation of the UNIX operating system, even when running on a RISC cpu with a 40MHz clock, can require upwards

---

[27] Simultaneity on a uniprocessor computer system implies that context switch and other times attributed to machine state maintenance are small compared with the frame times, T_f, of each process in the multiprocess structure.

of 100 μs or more for a context switch. So timeslicing 50 processes consumes 5 ms, which is a big penalty, and may even be prohibitive for the context of a real-time simulation approaching a 200 Hz dispatching rate.

Figure 4.4 illustrates an executive control structure for a three process simulation. The synchronous process, comprised of tasks labelled sync_task_{1-4} instanced in the synchronous table, is scheduled to execute more frequently than either of the two asynchronous processes. The scheduling of processes used here reflects a *rate monotonic* structure.

This scheduling algorithm is enforced when the platform's OS supports a preemptible kernel with multiple process priority scheduling levels. By assigning the synchronous process a priority level which is higher than the two asynchronous processes, the synchronous process is guaranteed to garner first rights for processor resource.

The primary distinction between the single process simulation discussed in section 4.3.1, and the multiprocess simulation structure outlined here is the occurrence of multiple (two in this case) asynchronous processes, each constructed from a unique task table. An asynchronous task, by definition, is scheduled to execute with a reduced frequency (hence the larger scheduling values, Δt_2 and Δt_3), and the time for a task's completion is less rigid (recall the strictness deadline criteria). This implies that tasks associated with both asynchronous processes may not necessarily complete before the next scheduling interval.

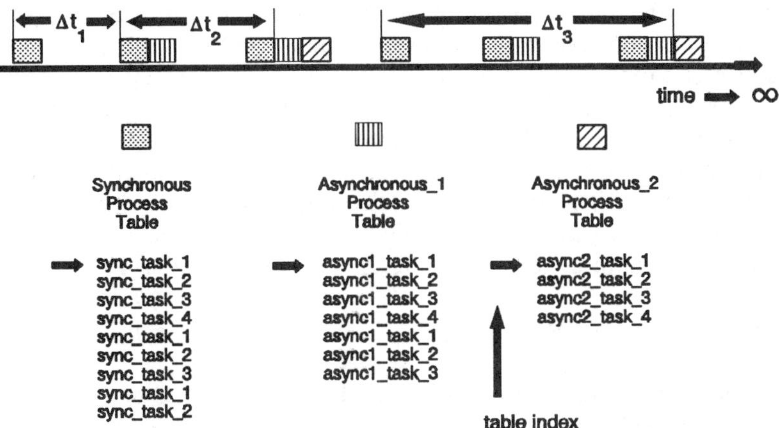

Asynchronous processes (1 & 2) are slaved to the arrival of the platform clock interrupt which is handled by the synchronous process. The executive control structure conducts scheduling for the simulation with a ratio of 3 : 2 : 1. For each 3 synchronous process contexts, 2 asynchronous contexts in Asynchronous_1 and 1 context in Asynchronous_2 are scheduled for execution.

Figure 4.4 Multiprocess executive control structure.

This situation is quite acceptable in most cases, as the asynchronous task context is preserved with each switch, and on return to the asynchronous process, execution continues at the point of suspension. The asynchronous task table indices are thus incremented in a lazy fashion relative to the synchronous process, as the executive structure permits the asynchronous processes to undergo multiple context switches during task execution.

Analysis of the dispatching intervals for the executive control structure of Figure 4.4 reveals that for each three synchronous contexts permitted to execute, two asynchronous contexts for asynchronous process #1, and one context for asynchronous process #2 occur.

In the case of the synchronous process, and tasks under its control, frame over-runs must be carefully monitored. A frame is said to *over-run*, or consume more time to complete its computations than are normally acceptable or expected, when a particular synchronous task within the task table body does not complete prior to the next index change in the task table index. Since each task within the synchronous process is allocated Δt_1 seconds to execute, an over-run will occur if, at the time of the next interrupt, Δt_{i+1}, the synchronous task executed at interrupt Δt_i is not complete.

Thus, if the quantity $| \Delta t_i - \tau_i | \neq C$, then a frame over-run has occurred. The constant C is a *boundedness* criterion. A synchronous task must not exceed a predefined time limit, and be considered as predictable. The specific simulation action to execute on this event varies with the application. Operator notification is one possible action in the case of non-life-threatening simulations, while some other more drastic recovery action may be needed in critical situations.

An over-riding goal of real-time simulation design is to achieve a situation where the processor resource consumed at run-time is less than the available amount. A surplus of cycles can always be filled by running bigger problems or adding more tasks to account for greater simulation fidelity and articulation. The converse of this situation is always unwelcome, like an unexpected dinner guest.

Tuning and performance monitoring of frame times can provide the necessary statistics for balancing the simulation. Balancing a simulation implies that task priorities are lowered or raised, or that limits are placed data structure sizes. Tuning and performance monitoring techniques are discussed in section 4.4.

4.3.3 Preemptible OS Kernels

Most multitasking OSs with preemptible kernels allocate processor resource to the processes with the highest priority until they either suspend for I/O or block on an external event (such as another timer interrupt or semaphore), and then the next lowest priority process ready to run is scheduled to execute.

A preemptive kernel is an essential feature which contributes to the qualification of an OS as suitable for hosting a real-time simulation. The real-time simulation must be unencumbered by any typical timesharing constructs such as swapping, spooled device processing (line printers), filesystem structure integrity checks, and other maintenance functions. These support processes interfere with the effective conduct of a real-time simulation, and a preemptive kernel permits these processes to be superseded by the executive control structure and the simulation processes created and controlled by it.

4.3.4 Interrupts

A popular mechanism for signalling a process to service a device or another process is through an interrupt service routine (ISR). An ISR acts on an external or internal interrupt signal by trapping the interrupt signal, causing a context switch of the current process, completing the associated function such as performing I/O or setting a semaphore, and then restoring the context of the interrupted process.

Interrupts can be either synchronous or asynchronous in character. The clock interrupt is a synchronous one, while external signals generated by devices or processes in response to preprogrammed events are asynchronous. An input from a touchscreen is a typical example: the operator touches the screen, the touchscreen controller signals the interrupt control logic configured to enable the ISR, and then reads the coordinates of the touch from the controller. The coordinates may be used by another process to determine the functionality associated with the portion of the affected screen. Asynchronous interrupts usually present interesting situations, for they occur almost at random, and thus require extra design and testing measures to ensure that they are correctly managed.

4.3.4.1 Polling and Asynchronous Interrupt Service Routines

The alternative to asynchronous interrupt processing is polling. A polling loop is established to constantly query a particular variable or input for changes. Polling is intrinsically more expensive than asynchronous interrupt processing through an ISR. A separate context is required for a polling process, and the process will do nothing except perform an indefinite loop until a change of state or action induces the polled quantity to change.

A polling process is an acceptable alternative to interrupt processing via an ISR when the interrupts are periodic and frequent. To see why, assume that the external interrupt frequency is ν_i, the polling loop frequency is λ, and further assume $\lambda \ll \nu_i$. When $\nu_i/\lambda \gg 1$, then interrupt processing can be considered as expensive, especially under the circumstance that each interrupt must be handled. It is cheaper, from a processor resource utilization perspective, to apply a polling process that runs asynchronously since fewer cycles are expended with this approach, although this option may not be feasible for a critical task, since the system may be compromised by this economical alternative design.

In the scenario above, multiple interrupts transpire during the time required to execute each polling loop, and lead to an *aliased* poll, which can be critical if the interrupt signal assumes a range of values, not just a discrete on/off level.

If the polling loop is faster than the interrupt frequency, and if each and every interrupt must be serviced, the polling option consumes more resource than does an interrupt handler. Ideally, $\nu_i/\lambda \geq \frac{1}{2}$ which implies that for each two cycles of the polling loop, one interrupt is supplied in perfect synchronization and phase, which also satisfies the Sampling Theorem criterion.

When $\nu_i/\lambda \ll 1$, an ISR becomes and even more attractive alternative, since interrupts occur less frequently than the polling loop cycle frequency (e.g., the interrupt acquires an asynchronous character). The predominant factors to consider for ISR mechanisms are the context switch time and interrupt latency period. The context switch time is the overhead associated with the preservation of the platform's state prior to entering the ISR body. The interrupt latency is the time interval between the external interrupt changing state or becoming active, and the initiation of ISR processing. This activity is illustrated in Figure 4.5.

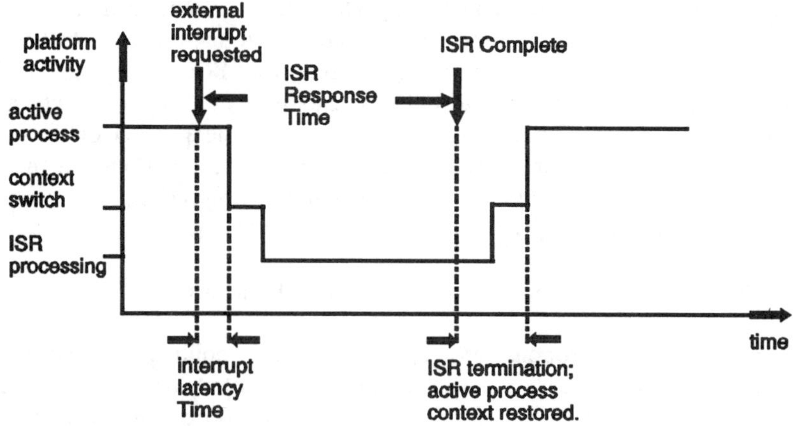

Figure 4.5 The temporal relationship between context switch time, interrupt latency, and ISR execution [Sav85] (Copyright © 1985 Van Nostrand Reinhold, with permission).

Platforms which possess micro-coded process schedulers typically require less than 10 μs to context switch, so interrupt latency is minimal. Context switches of this order are fast becoming typical for reduced instruction set computers (RISC). The time required to complete the ISR and restore a process context is the dead-time attributed to the ISR processing. A processor may be ill-behaved during this period, if it receives another interrupt signal. If the processor does possess interrupt inhibit capacities, caution should be exercised to exclude events which may force divergent behavior.

4.3.4.2 Vectored Interrupt Processing

The number of interrupt lines a processor possesses is limited, since interfaces to the external world outside the chip environment consume connector and packaging resources. One method of expanding limited interrupt processing interfaces is to use a vector-interrupt control mechanism. This method entails the encoding of an interrupt signal, a simple form of multiplexing. A binary to octal chip encoder with eight input lines, and three output lines is enough to encode eight discrete interrupt sequences into an octal format. This configuration is shown Figure 4.6.

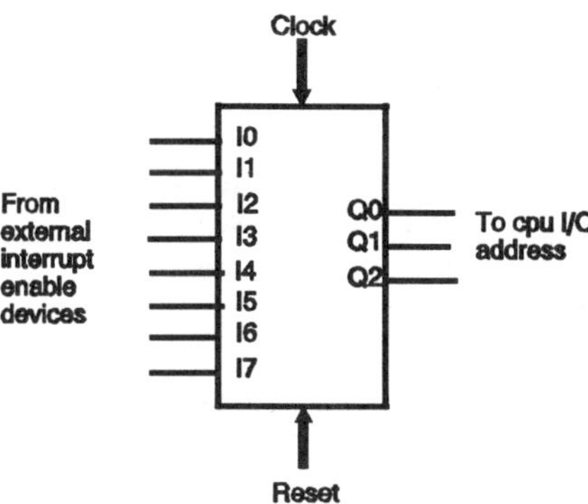

Figure 4.6 A binary to octal encoder for vector interrupt processing.

An external interrupt delivered to this encoder mechanism is translated into a 3 bit mask detected by the processor, usually through an i/o address. This mask value is read, and used as an index to an ISR table, a list of function addresses. The index is used to fire the appropriate ISR routine.

On many commercially available microprocessors, a *configuration block* is used to load interrupt vectors during bootstrap. The configuration block holds the address of interrupt routines that are triggered when its interrupt mask is accessed. Very low latency response ($< 50\,\mu s$) is possible with this structure. An image locater utility is often used to place the interrupt vector addresses, construct the configuration block, and map segments for data and code into memory prior to bootstrap. This procedure is particularly common for embedded microprocessor control systems.

4.3.4.3 Priority Interrupt Processing

Under certain circumstances, the prioritization of interrupts is essential to guarantee consistent and deterministic processing. Manufacturers of microprocessor chip sets usually supply programmable interrupt controller for this purpose. The Intel 8259 is commonly found in personal computers for mediating access between serial ports, disk interfaces, and memory.

4.3.5 I/O Operations

Real-time simulations often demand I/O functions and actions as a direct product of their execution. Whether the I/O manifests as a parallel or serial-digital stream, a snap-shot of memory stored to disk, or some other form of point-to-point information transmission, the I/O operation consumes substantial machine resources, since I/O is often the slowest platform attribute.

Table 4.2 classifies I/O operation by speed and method of transport. It shows that direct memory access (DMA) consumes the greatest bus bandwidth. The processor is idle during DMA cycles to preclude modification of the data written into the I/O subsystem during the transfer. Serial disk I/O is slower, since electromechanical storage systems require 10-20 ms to reach a given media track or sector before the operation commences.

Table 4.2 I/O method and relative speed classifications.

Comparison of I/O Speeds

Physical I/O Method	Relative Speed
Direct Memory Access (DMA)	Limited by memory bandwidth (300 Mbytes/s typical for a 64-bit bus)
FDDI or Ethernet	100 or 10 Mbits/s
Disk	5 Mbytes/s SCSI; up to 100 Mbytes/s for striped configurations
Serial (RS-xxx)	less than 100 kbaud

When disk I/O is conducted in a striped configuration, the bandwidth to and from memory can be substantial (see [Liv85], [Oga90], and [Cab91]). Striping depends on the precise synchronization of disk platter rotation. Information is spread across multiple disk platters residing in physically distinct disk drive units, and is retrieved in an identical fashion. Rather than a 5 Mbyte/s bandwidth between memory and media, a striped disk configuration with N disk units in the strip can deliver up to N times the standard disk I/O rate, providing drastically better speed.

Disk striping is predominantly supported by platform vendors who provide symmetric multiprocessing options, as these machines possess I/O subsystems designed to supply a high-speed memory interface. Striping disks in a run-of-the-mill platform, like a low-end personal computer, will produce dubious benefits, since the memory bandwidth in these platforms is quite low.

It is interesting to note that many shared-memory supercomputers offer striped disk configurations, and many installations run them, but loading a multigigabyte (8 gigabytes) dataset can still require many hours to accomplish [Cor91b]. This suggests that shared-memory systems are intrinsically unbalanced with respect to computation and I/O bandwidth.

In the course of a real-time simulation, the state of the simulation may be saved (a copy of the global database is flushed to disk), for the purpose of post-analysis playback and debriefing. This operation must be accomplished without over-taxing the I/O subsystem, and is often conducted in an asynchronous process context. The rate at which the flushing operation occurs is dependent on the database size (the number of bytes to write on the disk), and the simulation playback fidelity requirement.

Suppose that a simulation platform is equipped with one disk I/O channel, and it is a 5 Mbyte/s stream, and a simulation requirement dictates that twice each second, the database must be stored to disk. These snapshots will be used during playback for post-simulation analysis. Obviously, the 2 Hz goal can be satisfied provided that the global database (or that portion of the database necessary to facilitate playback) does not exceed about 2.5 Mbytes. A slightly smaller amount is usually preferable, owing to overhead associated with I/O operations, but can usually be neglected if an asynchronous (non-blocking) I/O mechanism is supported.

Time is also consumed by scatter/gather operations. These functions are often performed by intelligent DMA controllers or I/O devices. Addresses are passed to a subsystem which must obtain the data from memory by a DMA transfer. During this time, processors cannot generally access the address and databus, thus robbing the processor of valuable cycles, while "memory is moved."[28]

I/O operations normally cause a process to suspend -- *block* -- until complete. I/O operations that are asynchronous return immediately from their invocation; they are *non-blocking*. Non-blocking I/O operations may be expedited by an ancillary control process (ACP), which is an autonomous agent executing on behalf of the calling process. ACPs are frequently associated with intelligent controllers for disks and network I/O devices. The ACP copies the contents of an address to the I/O channel or the device controller where the information is copied into the device (e.g., disk or network).

4.3.6 Interprocess Communication

Multiprocess simulations frequently depend on OS functions to support communication between tasks residing within separate contexts. Interprocess communication (IPC) is the term describing this class of communication support provided by an OS. Under UNIX for instance, IPC is facilitated by pipes, shared-memory, message-queues, and sockets. These functions provide a vehicle to implement bi-directional communication between one or more processes which are either remotely based (in another machine

---

[28] As a general rule, computers do not move memory efficiently. They are built for arithmetic purposes. Message-passing parallel computers, multicomputers or hypercubes, are extremely efficient at moving memory. Block I/O transfers are comparably inefficient in uniprocessor or multiprocessor shared-memory systems.

or physical address space), or within the same address space.

In real-time simulations, the queue structure serves as a common mechanism for information exchange between processes. Queues are usually designated as last-in first-out (LIFO) or first-in first-out (FIFO). A LIFO queue is read from the top: the elements inserted into the LIFO queue are stacked up like boxes in a warehouse. The top boxes are stacked last, but removed first when inventory is reduced. Conversely, the FIFO queue permits element removal in an arrival-first order, from the bottom up as it were.

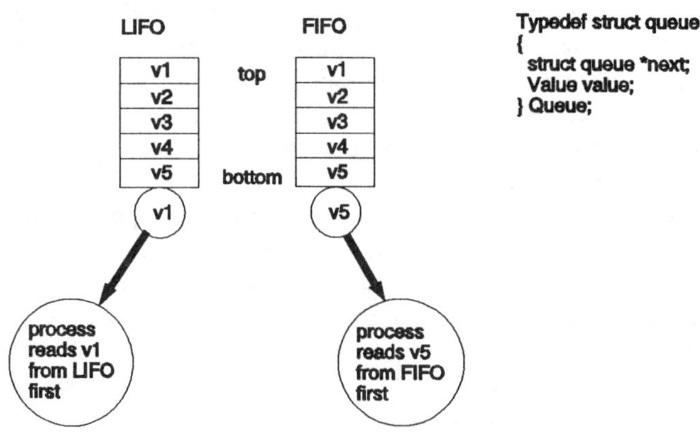

Figure 4.7 Queue structure and operational modes.

Figure 4.7 illustrates the two types of queues, and their basic operational modes. The data structure for either queue type is the same. Queues are a form of recursive data structure; they can be traversed like a linked list, where each data element attached to a pointer contains the value either to read or write. Queues are very simple and plain structures, and they are also easy to implement. In the real-time environment, queues are usually statically allocated and have a finite extent, since memory allocation in real-time is not always predictable.

Two processes communicating via a shared-queue structure, in the multiprocess/multiprocessor environment should use a semaphore to guard against corruption of the stacked value during either queue-write or queue-read operations. The functions implementing the queue insert/remove operations should include access exclusion via a semaphore. When a process attempts to either write or read from a queue by invoking a queue insert/remove function, the semaphore must be obtained first. A process will block on the semaphore function until the process which holds the semaphore releases it. The exclusionary capability of semaphores minimizes contention and prevents corruption of data. Uniprocessor systems have no outstanding requirement for semaphores, since each task and process within a uniprocessor system executes sequentially. Contention and corruption is often a non-issue for uniprocessor systems.

Because the queue depth is statically allocated, the queue insert/remove functions should report a status return value indicating success or error. Status values for a queue-is-full error code (for insertion), a queue-is-empty code (for removal), and a queue-op-successful are usually sufficient for most applications.

4.4 Performance Measurement Assessment

The finest sense of accomplishment and satisfaction resulting from a real-time systems engineering effort arises from the verification of the system's performance fidelity, the match between requirements and the final product's run-time capability. A system's run-time fidelity is a principle element of quality, but not the sole determinant in the quality equation. Safety, reliability, maintainability, reusability, and other parameters must receive equal consideration not only during assessment and testing, but throughout the requirements and design stages of a real-time product.

The novice engineer who bristles with pride at the conclusion of acceptance testing over the simulation's fidelity may not fully recognize the importance of tangential or collateral engineering issues. The cloak of technological complexity and accomplishment can easily shield a persona from the crucial safety and reliability issues intrinsic to a real-time system. Awareness of these subjects must be inculcated and learned through training, practice, and membership of professional organizations. Quality issues and factors affecting collateral systems safety are better acquired on the job. Often, the only way for an engineer to appreciate the importance of safety factors is for him or her to gain first-hand experience with the consequences of a failure. This happens far too often (see Chapter 5 on Software Safety).

System performance is the most visible element representative of a product's quality. Reliability issues generally materialize in the maintenance cycle, where frequent repair or downtime serve as harbingers of poor design, or failure to successfully implement requirements. In real-time systems, performance is closely coupled to the execution rate of the simulation, and the realism generated by the computational representation of modeled phenomena.

4.4.1 Simulation Performance Measurement

A simulation's fidelity results from the successful interplay of two ingredients: the temporal and physical reconstruction of characteristics and behaviors found in natural and man-made systems, which are difficult or too expensive to recreate or model via material methods. For example, one does not build an entire skyscraper and subject it to earth tremors to assess mechanical structural stresses. Instead, a finite element model is used to find material deformations, or a scale model is built and placed on a shaker table with accelerometer instrumentation to measure beam displacement.

The performance attributes of a real-time simulation product are first estimated during an early phase of product development, tuned during integration of the software, and verified by testing under known conditions. Test conditions which mirror those likely to be encountered in the field are essential to verify realistic performance and quality objectives.

Hardware system performance, especially of the I/O subsystem, is generally easier to verify. The hardware configurations do not change with time, unless they are designed with fault tolerance and redundancy in mind. Thus, checking to see if an I/O channel delivers a data stream at a required rate implies that the channel is turned on, and the emitted stream is examined with an oscilloscope or a counter to detect the rate of transmitted data.

A computation element, like a microprocessor, is a bit more subjective. Various benchmark codes for evaluating floating point performance, integer performance, and OS performance (among others), all provide some indication as to the effectiveness of a microprocessor for certain attributes associated with commonly occurring algorithmic structures. Unless the simulation software has been prepared and installed on the microprocessor, a true measurement of performance, in terms of millions of instructions per second (MIPS), or millions of floating point operations per second (Mflops) cannot be objectively decided.

One approach to predicting the time required to execute a block of code involves the counting of instructions, at the assembly language level, emitted by the compiler or code generator. This technique assures an accurate measurement, since the cycles required to address operands and execute the opcode are known from a microprocessor's architecture. An automated tool for this activity has been implemented and tested for a small number of microprocessor architectures (see Park and Shaw [Par91]).

A common approach for timing code segments is to insert calls to real-time clock/timers at the beginning and terminal phases of a segment, and then subtracting the values returned from the two timer calls. This technique is useful for measuring subroutine invocation overhead, the cost of creating a call stack-frame, which loads the arguments of a subroutine onto an area of memory reserved for this purpose, and passes control to the subroutine by advancing the program counter to the appropriate instruction stream.

A less common method involves an oscilloscope and infinite loops (see Razouk *et al.* [Raz86]). One places an infinite loop around a code segment, and inserts a statement at either the terminus or beginning of the loop (and internal to it) to assign a value to a specific address in the memory space or to set a bit in the processor's status word. Then connect an oscilloscope probe to this address or status word bit position, and examine the oscilloscope display as the loop executes. The display will show a sequence of spikes or impulses matching the precise times that the loop writes to the address or status bit.

A hardware monitor of this form will produce unambiguous timing information, but it is an invasive procedure, and one of last resort especially during integration and testing, unless the procedures at these phases specifically mandate their use. An equivalent, but far more convenient method is the software monitor discussed below (see section 4.4.4 on Monitoring).

If the hardware monitor device is replaced with a logic analyzer accepting multiple input lines equal to the word size of the microprocessor and connected to the memory address associated with the desired variable to monitor, then the datum's contents can be ascertained along with the time it was written or last modified. This method is quite slow for verifying the correctness of each variable used in the simulation, since a lot of labor is expended connecting to different physical addresses on the memory card.

4.4.2 Tuning

Subroutine invocation overhead can be quite costly. The author recalls one project that required a function to test if two entities were visible to each other across a gaming area containing mountains, hills, and other obstructions. The clear-line-of-sight (CLOS) determination depended on evaluating terrain elevation levels relative to each entity in the gaming area. This operation was not "terribly" expensive to conduct, requiring about 45 μs for each pair of entities.

But the nature of the simulation was that many pairs (the simulation requirement dictated 300 LOS-pair evaluations/s) had to be tested for CLOS determination. When considered in isolation, 45 μs is a pittance, but since the CLOS process was one of ten real-time processes competing for processor resources, the problem became severely magnified.

The host computer consumed 12.7 μs for subroutine overhead each time the CLOS terrain evaluation function was invoked. The terrain evaluation function returned the altitude as a function of x and y. The subroutine overhead amount to 28.2% of the cumulative wall clock time for each evaluation, and is wasted by the machine preparing call frame-stacks. The simplest way to recover this time, or a large majority of it, is to eliminate the function invocation call altogether, and substitute an in-line expansion macro (in C) or statement function (in FORTRAN) for the routine. The in-line expansion macro was selected, and 95% of the time consumed by the subroutine invocation overhead was reused by the CLOS computation, thus reducing each CLOS evaluation to under 36 μs. This simple realization produced a CLOS processing rate which exceeded the simulation requirements.

This example illustrates the memory versus speed tradeoff common to all software engineering efforts. On one hand, memory constraints, as well as programming style, preclude a complete in-line algorithmic construction. While eliminating subroutine overhead, the software consumes more memory since code is continually replicated, but the execution speed improves.

Other techniques for tuning simulations include the elimination of computation where it is unnecessary. This implies that conditional tests should precede an expensive code segment, since microprocessor-based conditional evaluations are many times faster than any floating point scalar computation.

When task frames over-run predefined specification limits, the frame is said to be *unbalanced*. There are no sure-fire remedies in these instances. Perhaps if some tasks are moved to another process, or the number of task instances is reduced in the frame (meaning that the rate of execution drops for the task). Also, reducing the amount of work enqueued at any time certainly reduces frame execution time (T_f), and raises dead-time. Restricting queue depth may help, but may also constrain and reduce the simulation fidelity by preventing enough real-time results to emulate the natural phenomenon.

4.4.3 Debugging

The debugging process is a complicated affair in a real-time scenario, since the time element shares the spotlight with the logical computational results. In many cases, the computational results for a high-level language simulation of an algorithm can be debugged and investigated under the relaxed environment of a symbolic debugger. The symbolic debugger permits single-step examination of source code operation, along with variable content and address examination.

Provided that the algorithm possesses a spatial (logical) dependence which can be decoupled from the temporal algorithmic process, a symbolic debugger serves as an expedient vehicle for checking the conditional and numerical processing of the algorithm. Abstracting an algorithm which is both logically and temporally dependent -- one characterized as a function of logic and time, i.e., $f(l,t)$ -- into a form which separates the two independent variables into a product, e.g., $f(l,t) = f(l) \bullet g(t)$, produces a relaxation in the testing requirements, so far as the logical computation aspects are concerned.

Symbolic debuggers are acceptable mechanisms for investigating anomalies within the logical simulation structure, the conglomerated superposition of functions, processes, data, and algorithms used to build the simulation software. This means that all conditional and non-temporal aspects of the simulation can be checked and explored without distorting the logical results.

The temporal context is not so fortunate, however. Symbolic debuggers add notoriously large amounts of overhead to a context, the by-product of extending symbolic observation and breakpoint operators to the software. Breakpoint symbolic debuggers do not preserve the temporal simulation structure; they are poor instruments for evaluating timing characteristics and other temporal aspects of simulation fidelity.

4.4.4 Monitoring

The hardware monitoring scheme previously discussed is quite laborious and not very flexible. In general, hardware monitors should be avoided for the purpose of software verification due to the expense and time involved in their implementation. An alternative to the hardware monitor, one that is particularly effective for real-time systems, is the real-time symbolic software monitor (RSM).

The RSM utilizes the flat symbol table generated by the DDE as a template for the organization of memory specifying the variables used by the simulation. Flat symbol tables consist of simple types: real, double precision, integer, Boolean, and character elements. They may contain matrices of these types as well. DDEs can provide a more robust implementation that permits articulated data structures and unions.[29]

The RSM is driven from a commandline grammar, which may be constructed from standard compiler construction tools. The grammar should be simple but

[29] A C++ class should be built for this purpose.

functional. The two most common syntactical elements for the monitor grammar are "poke" and "peek" operations. The poke operation permits a logical address within the symbol table to be set to a given value, while the peek operation permits continued observation of the logical address contents, in real-time. Additional grammar elements may include the notion of a "gate" or "trigger." These operators enable snapshots of symbolic addresses to occur in a specific order, relative to the Boolean outcome of an addition monitor statement. For example, the grammar *peek t_up(1,1) @ 10 Hz when v_xyz < -1.2* commands the RSM to begin monitoring the address **t_up(1,1)** after *v_xyz* assumes a certain range. The RSM should also support a grammar queue, where multiple groups of monitoring commands can be loaded, initialized, and activated as needed.

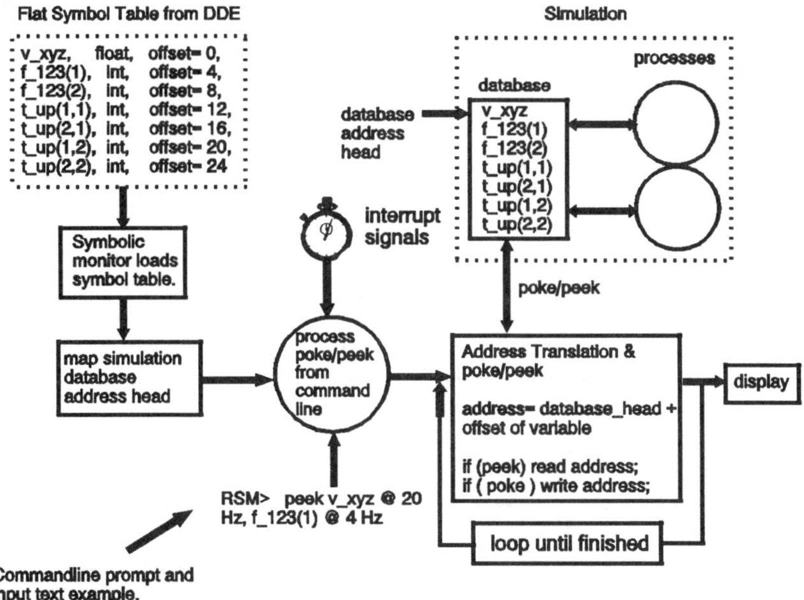

Figure 4.8 The schematic of a symbolic monitor for the inspection and modification of real-time process variables.

Figure 4.8 shows a schematic of the real-time symbolic software monitor in an operational context. The initialization phase processes the flat symbol table, and obtains the database's starting address in memory following the simulation's initialization.

The commandline interpreter processes the input grammar. In this case, two variables within the database are to be monitored at different rates. The symbolic software monitor incorporates a timing mechanism similar to that used for the simulation execution control structure. Thus, the variable **v_xyz** is examined ("peek") 20 times each second, while **f_123(1)** is examined 4 times each second.

The symbolic monitor can be a powerful tool for assisting with acceptance test procedures, since this software mechanism is far easier to implement for a class or range of variables than the hardware monitor system.

4.5 Considerations for Testing Real-time Systems

Testing a real-time system, and verifying that performance and other critical requirements are satisfied raises many challenges. The acceptance test plan (ATP) is one of the key documents in the software lifecycle, for it specifies how the system will behave, and under what conditions a preconceived state and known result will be achieved. The ATP calls out how the system will be tested, and if the tests are passed, then the customer should purchase the product based on these preconceived and illustrative cases of product performance and exercise.

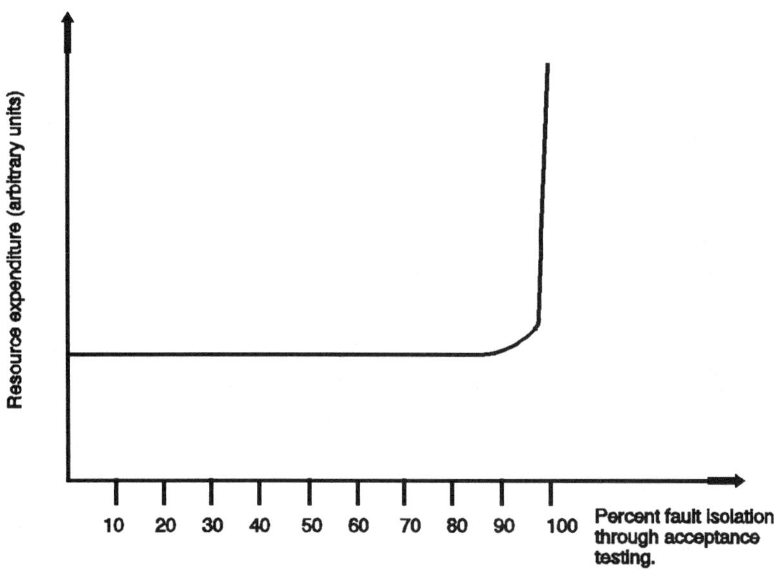

Figure 4.9 Resource expenditure requirements to obtain a percentage of fault isolation certainty in a complex system.

A test plan should be built by at least one senior individual, one who has held primary technical responsibility for the project from its conception. For complete neutrality in ATP execution, an independent agency or organization should conduct the test plan, and record discrepancies in log books. The independent agency, often the quality assurance group or test operations personnel, hold preeminent responsibility for acknowledging whether the product satisfies the test plan goals and requirements, and is thus qualified for shipping.

Once a product is shipped to a customer, liability for performance in the initial phases of field testing usually lies with the contracting firm. Thus, it is imperative to verify that the product is incapable of adverse performance which may jeopardize human life or property. To secure this level of quality and safety, the product must undergo scrupulous evaluation prior to release and shipping.

What constitutes "scrupulous evaluation?" Sadly, and especially of late within the United States National Aeronautics and Space Administration (NASA), satisfactory testing of critical mission components and systems are sacrificed because of budgetary constraints [Brd91]. Testing requirements must be driven by the operational and functional objectives needed to sustain a successful and reliable implementation.

The ATP organized for this purpose must strike a balance between delivery schedule, test expenditures, and safe operational performance at a nominal level. To assure operational performance at a nominal level, the entire system must undergo a collective exercise to detect integration flaws not evident at earlier or lower-level component testing stages.

The cost of testing a complex system, in an effort to locate potential failures, obeys an exponential law, as greater certainty in fault isolation is desired. Figure 4.9 illustrates this dependency. Thus, it is nearly impossible to explore all possible combinations of initial and runtime conditions during ATP applications, as the cost is prohibitively expensive.

Concluding Remarks

By providing this overview of real-time system properties, design elements, and methods for debugging, I have attempted to explain the basic concepts and ideas used to build and implement this important class of simulation on sequential and shared-memory systems. A complete discussion involving detailed hardware designs, or esoteric executive control structures for multiprocess simulations is beyond the scope of this text, and unnecessary for the purpose at hand. It is the author's intent to describe and impress this background material on the reader for preparatory, but not religious purpose. Each real-time simulation I have seen is distinct, separate, and entirely unique from each other. The discussion given here is only a indication of the variety which real-time systems design and function can demonstrate.

Suggested Reading

Stephen Savitzky has written a fine introductory text on real-time systems [Sav85]. Hatley and Pirbhai [Hat88] provides a rigorous introduction to specification and design for real-time systems according to popularly practiced methods. Heath [Hea91] has written a fine text which details the internal design for a multitasking operating system. It is a terrific case study with timing data and analysis of process scheduling and performance. Kalbfleisch, *et al.* [Kal91] have recently benchmarked five of the most popular real-time operating systems available from third-party vendors. An excellent tutorial on topical issues for hard real-time systems research, design, and implementation is the compendium by Stankovic and Ramamritham [Sta88]. Welch [Wel87] describes an approach for managing hard real-time constraints on a transputer. Jain [Jai91] has compiled a large collection of performance measurement, tuning, benchmarking, and evaluation tools and techniques. Over 150 case studies are included in this text.

A detailed outline of software documentation requirements and lifecycle activities can be found in DoD-Std-2167A, the Department of Defense document which covers defense-related software development activities [DoD85]. Roetzheim's text ([Roe91]) describes the mechanics and procedures for conducting a software lifecycle according to DoD-2167. An equivalent Department of Defense Standard, one promulgated by the Navy is DoD-Std-1679A [DoD83]. The ANSI/IEEE Software Engineering Standards can be obtained from the Institute of Electronic and Electrical Engineers [IEE83].

5

Software Safety

This chapter presents an introduction to software safety issues for real-time systems. The approach to software safety used here presents elements of traditional analysis techniques such as Petri nets and software fault trees. The discussion then turns to formal specification methods. Formal methods are generally believed to provide the most precise and expedient representation of software requirements and design, and these attributes underpin the foundation of safe software systems engineering.

5.1 Definitions

Software safety is an attribute of product and/or system quality. System reliability has traditionally served as the principle product quality measure. But this metric has become obsolete as a primary determinant for software quality. Software-based systems and components deserve an articulation of quality beyond standard reliability measurements. Reliability focuses on the ability of a system to objectively perform and meet specification. Engineering decisions, implemented during requirements and design, guide a system toward correct operation. "Reliability theory currently tries to quantify errors, safety implies that errors must be qualified." [Lev81]

A computer program may continue to run, even though incorrect computations take place during execution. Reliability techniques generally provide information on failure intensity, not the characteristics of failure. A *failure* is defined as a system's inability to perform a preordained function within specified limits.

Leveson [Lev83] classifies a *safety failure* as one which leads to "casualties or serious consequences." Obviously, safety failure includes death or injuries, while 'serious consequences' also implies loss of property or tangible assets. Safety failures have outcomes of this type, while non-safety failures do not.

An ATM system that will not transfer funds between accounts at the command of the operator is an example of a non-safety failure. This is an inconvenience, not a hazardous situation. To remedy a non-safety failure, the system may be reconfigured by a revision to the keypad code controlling ATM electronic fund transfer.

Safety failures involve critical system functions, where reconfiguration is not always possible or expedient. Execution time constraints, or the unavailability of redundant components may preclude failure recovery.

Safety and non-safety failures result from *faults* embedded within a system. A fault is an error. A fault may be *latent*, in which case the failure may not appear for a long period of system operation. Leveson [Lev83] states, "If recovery is not possible after a safety failure, then fail-soft or fail-safe procedures must be activated."

Fail-soft systems can continue execution, with diminished or reduced functionality, after a safety failure is incurred, until such time as the failure is fixed. Fail-soft systems attempt automated recovery from failures, and are intrinsically more expensive and complex. Interplanetary spacecraft are well known for their fail-soft capability.[30]

Fail-safe systems attempt to minimize damage or losses by terminating programmed operation and aborting. Fail-safe systems possess the capacity to prevent safety failures. Fail-safe systems might appear in ICBMs to prevent warheads from accidentally arming, thus limiting the potential damage that can be caused.

> "The goal of software safety is the avoidance of system safety failures which are caused by a software error and/or are detected and handled by software procedures. A *safe system* is one which prevents unsafe states from producing safety failures where an *unsafe state* is defined as a state which may lead to a safety failure unless some specific action is taken to avert it. In order to prevent safety failures, it will be necessary to be able to detect unsafe states." [Lev83]

The detection and diagnosis of unsafe states, through manual or automated procedures, during design, testing, and operation is an active field of research and investigation. Nuclear power plants could be safer to operate with a real-time safety analysis system which might detect the emergence of unsafe conditions prior to the evolution of a serious event or safety failure. Telephone and computer networks may be organized with a diminished probability of failure and downtime if a sufficient mechanism for software safety assessment detected the likelihood of failure.

These goals and applications remain elusive. But safety failures receive growing attention, as systems dependent on software are progressively infused into society and are used daily. The expense required to construct more complex systems lends motivation to government, industry, and private concerns to search out alternative mechanisms to lower costs and improve software safety. Safety is concerned with the consequences of failure.

5.1.1 Software Reliability

When a computer program does not perform or execute correctly, a software failure prevents satisfaction of a preexisting specification. The source of the failure is a fault

---

[30] NASA's Voyager II spacecraft developed a hardware fault at one address in the region reserved for image recording on approach to the planet Uranus in 1986. Had the hardware fault occurred in the attitude control RAM, a patch may have been impossible, as the spacecraft could have tumbled out of control.

in the software [Mus89].[31] A software failure is due solely to the malfunction of a software-related function or routine, and usually originates at the requirements, design, or coding phase of a lifecycle.

Faults can also be introduced during the maintenance phase. These semantic distinctions are important for assessing and classifying a software failure. A software engineer can isolate a fault to a few lines or modules of software, but it is the nature and characteristic of the failure which provides the clue or key to the fault source.

An important quantity for software reliability considerations is the fault intensity, the number of faults detected per hour of system operation. According to Musa [Mus89], software reliability is defined as "the probability of failure-free operation of a computer program over a specific period of time, and is an exponential function of the negative product of failure intensity and time."

For example, if a computer program exhibits a total of 6 faults over a 2000 hour period, its reliability figure (ρ) for any 8 hour period of operation is given as:

$$I_f = \frac{f}{T_e}\ , \textit{ where } f \textit{ is the \# of faults observed over a execution interval } T_e;\quad \rho = (1 - I_f * T_e);\quad \textit{So, } I_f = \frac{6}{2000}, \textit{ yielding} \quad \rho = 1 - (.003*8) = .976;\quad \textit{over } T_e = 8 \textit{ hours.} \tag{5.1}$$

The equation states that the computer program will reliably operate, in a probabilistic sense, with a 97.6% success rate over any given 8 hour execution interval, T_e.

The notion of time is critical to the reliability assessment. Since a computer program may execute for extended periods, an individual module may only become active during a small percentage of the execution interval. Thus, the reliability calculation for any module must be calculated according to execution interval, not the passage of calendar time which assumes a 24 hour day.

Calendar time is an expedient measurement of project schedule and program performance. But reliability measurement is initially based on execution time, so a conversion is needed. We use [Mus89] to obtain a definition for the conversion, who states:

> "The instantaneous ratio of calendar time to execution time is the resource usage rate for the limiting resource at the time, divided by the available amount of that resource. The resource usage rate is the rate at which resources, such as person-hours, are expended as a function of the execution time used and the number of failures."

---

[31] We assume that the underlying hardware is fault-free. Hardware fault-tolerance schemes are used to ensure reliable operation of sensors, cpus, controls, etc.

If the period of computer program development is limited by test personnel, and suppose that 4.5 worker-hours are needed to repair each failure, which occur with an I_f (the so-called failure intensity) of 12 per cpu-hour, then the resource usage rate is 54 worker-hours per cpu hour. If only 2 test operators are available, the instantaneous ratio of calendar time to execution time is 27 worker-hours per cpu-hour.

The basic execution-time model presented and discussed here is derived with the assumption that the software failures can be modeled as a homogeneous Poisson process in execution-time. This means that the failures observed within a specified time period obey a statistical pattern -- the Poisson distribution. The Poisson distribution produces a mean value and deviations for I_f over the execution interval.

Several software reliability models have been created and tested on a variety of projects, ranging from the Space Shuttle Mission Simulator to A.T.&T.'s 5ESS telephone switching system. Software reliability models are useful for predicting field maintenance requirements, which can in turn be used to estimate the need for service personnel staffing. A thorough treatment of software reliability is given by Musa [Mus87].

Software reliability measurement and assessment is useful to predict and monitor average software failure rates, but the models are virtually incapable of pinpointing the potential for catastrophe which might arise from a fault-induced failure. Hardware failures are often isolated to single components, while software failures may involve several modules or processes, and possess a high degree of correlation. Traditional hardware-based reliability calculations, when applied to software, may produce specious results [Lev83].

Rigorous -- almost infinite -- testing and resource expenditures are needed to preclude a catastrophic failure for a safety-critical software system. No organizations can commit infinite resources to attain a perfect reliability figure of merit ($\rho = 1$). A cost and safety tradeoff must be made which ultimately sacrifices the reliability of the finished product.

5.1.2 Examples of Software Failure

If a software-controlled system fails, such as a power plant, telephone switching system, x-ray machine, or pacemaker, monetary and human losses can result. Awareness of system safety, methods for assuring reliable operation, and the controlled functional degradation of system performance, through either fail-soft or fail-safe technique are individual and corporate responsibilities of ethical importance.

Despite individual or organizational efforts to eliminate and preclude software failures, they undoubtedly arise, as complex systems require prohibitively expensive measures to exhaustively analyze and test. Errors, faults, and the failures they can induce might slip through the testing and evaluation process. Serious consequences may result if latent faults materialize during system operation.

The rapid computerization of national industrial and economic infrastructure has proved extraordinarily successful, despite the rare failure which grabs a headline. The nominal operation and performance of the machines we use throughout our daily lives promotes continued confidence, and fosters a growing dependency.[32] The technological thirst for and consumption of improved (faster and more functional) electronic systems has an attendant danger in that each of our lives is exposed to greater cumulative probabilistic failure of these systems. But technological progress significantly contributes to economic growth and quality of life.

In the medical profession, computer-assisted tomography (CAT) scanners, ultrasound imaging machines, and other advanced diagnostic units are made possible by embedded microprocessor systems which control magnetic fields, x-ray dosages, and the processing of clinical data. With the microprocessor and software, a powerful team emerges which tirelessly yields improved medical service and care.

Advanced diagnostic tools save lives by delivering faster and more definite diagnosis. These machines justifiably deserve continued use and development within medical environs. Adjustment to and acceptance of advanced technology brings comfort and confidence, a trusted ubiquity earned through repeated performance and functional success.

But when a trusted system produces unexpected results or fails to operate correctly, unsuspecting emotional dependency is jarred. At best, mere inconvenience forces the consumer to search for an alternative, should one exist. Nuisances in this instance result in lost time and commerce, so a small monetary loss is calculated. An example of this nuisance is the departure delay of an aircraft due to the underestimation of take-off slots for a given day's traffic by aircraft routing software. Cargo and passengers sit on the ground, rather than moving closer to a destination. The alternative, death or physical injury, is far more serious.

Therac 25[33]

Although scarce and infrequent, when these instances of failure arise, they serve as embarrassing reminders of the fallibility in human engineered systems. Consider what can happen when an x-ray machine used to deliver radiation therapy for breast cancer patients delivers an overdose.

> "Mr. Bearden: ...After doctors removed a cancerous tumor from her breast, they used a state of the art machine called a Therac 25 to perform radiation therapy. But the software that controlled the machine had a [flaw]. Ms. Yarbrough says she knew something was wrong immediately.
>
> "Katy Yarbrough: I just sat up on the side of the table and she said, what's the matter, and I said, you burned me. And they said, oh, no, that's not possible, this is the most sophisticated piece of equipment on the market, and it's not

---

[32] For a soberingly humorous essay on this dependency, the reader is encouraged to read The Open Channel in the Sept. 1991 issue of *IEEE Computer*.

[33] Unless noted, all quoted text for the Therac-25 failure is stated in [Bea90]. Courtesy of MacNeil/Lehrer Newshour.

possible for you to be burned. And a week later I was totally paralyzed on the left side.

"Mr. Bearden: Like all software, the program that controlled the Therac 25 was composed of thousands of lines of instructions called codes. When a certain rare sequence of events occurred, the code acted in a way the software designer hadn't anticipated. Mrs. Yarbrough says she was supposed to have received 200 rads of radiation; she got 20,000 instead, that in spite of the fact that the Therac 25 had been used tens of thousands of times before and had never hurt anybody.

"Ms. Yarbrough: They denied that the accident even existed.

"Mr. Bearden: But you had physical evidence. You had a burn mark on your chest.

"Ms. Yarbrough: It actually became a hole. I had a hole all the way through my body.

"Mr. Bearden: And they still said the machine couldn't have done that?

"Ms. Yarbrough: Yes. They really believed the machine was infallible. It was that kind of a thing like it's just not possible, there's no way, and when you say it's not possible, and then you've got a person that has a perfectly normal arm and it has absolutely no value to me. We've talked about taking the arm off. I'd be no better with it off than I am with it on. It's not going to help the pain.

"Mr. Bearden: Mrs. Yarbrough said the treatment center continued to deny the machine had malfunctioned until the accident was repeated in Tyler, Texas. Thirty-three year old Ray Cox was getting his ninth treatment for a tumor on his back when he felt a severe jolt of heat. The machine was inspected. The technicians couldn't find anything wrong. It was put back in service. Twenty-two days later at this same center 66 year old Vernon Kid screamed in pain during his 30th treatment. This time technicians found the answer. Excessive radiation was delivered if an operator made a typing mistake while programming the treatment sequence and then corrected it in a certain specific way. Both Cox and Kid died from radiation exposure. The Texas machine was never used again but Therac 25 machines are in use in other states with new software."

In the past, hardware restraints and physical limit checking would have prevented the overdose (a knob or lever adjustment). But modern therapeutic equipment is highly programmable, and the old-fashioned fail-safe tolerance elements and circuit-breakers have been replaced with software-based Boolean conditions. That an operator could circumvent Boolean safety checks illustrates the increased vulnerability of software to be corrupted by unexpected or erroneous input. Human fallibility, both in operation and software design, combined to cause a serious failure.

Traffic Alert and Collision Avoidance System (TCAS)[34]

TCAS warns airline pilots about impending midair collisions, and suggests evasive action to avoid mishap in crowded skies. After 30 years of development, the TCAS has been initially installed in about 700 large commercial airliners. Plans call for the installation of TCAS in 4000 airplanes by 1993. But on May 2, 1991 the Federal Aviation Administration (FAA) ordered 200 of the 700 units, which cost US $150,000 each, removed from service.

These units, built by the Collins Defense Communications Division of Rockwell International (Dallas, TX), were generating fallacious collision signals. They detected aircraft which did not exist, and were instructing pilots to avoid incoming aircraft which were not there.

> "The cause was quickly identified as a software glitch. More precisely, it was a software gap -- five lines of code missing from the faulty units.
>
> "The problem arose in the course of testing, because Collins engineers had temporarily disabled the program's range correlation function -- a few brief lines that compare a transponder's current response with previous ones and discard any intended for other aircraft. Without this filter, the system can misinterpret a response as coming from a fast-approaching airplane.
>
> "After testing the systems, Collins shipped them to airline customers without re-enabling the range correlation. For the most part, the systems worked as intended. But in high-traffic areas where many airplanes are interrogating each other -- around Chicago, Dallas, and Los Angeles, particularly -- ghosts appeared frequently. Pilots were misled, and air traffic controllers were distracted from their routine tasks by the need to handle nonexistent situations.
>
> "In the range correlation scheme, the system notes the distance at which it first receives a response from another aircraft -- say 10 miles. At the next interrogation, the distance may be 9.5 miles. The system would then expect the next response to be at approximately 9 miles, and would set a range gate so that it could look for a signal at that distance and calculate the closure rate. Without this correlation, the system becomes confused."

The manager at Collins stated, "We had a simple human error where an

---

[34] Unless noted, all quoted text for the TCAS failure is stated in [Wat91]. All rights reserved © 1991 IEEE. With permission. See also Cushman [Cush91] for an update.

engineer misclassified the changes in the software...It didn't show up in our testing because one of the essential elements was absent; you have to have many, many TCAS-equipped airplanes in the sky."

Collins has adjusted its software quality assurance procedures to require that a committee of software engineers reviews changes prior to releasing a new product. The other TCAS manufacturers, the Bendix/King Division of Allied Signal, Inc. (Baltimore, MD), and Honeywell, Inc. (Phoenix, AZ) experienced no such failure with their units.

A.T.&T. Long Distance, Jan. 15 1990[35]

The U.S. telephone network is the most advanced and reliable in the world. Each day, long distance network circuits conduct about 80 million transactions. About 70% of this market is controlled by A.T.&T. But on Jan. 15, 1990, a fault in the out-of-band signalling software caused the network to enter a chaotic state reminiscent of an epileptic seizure. "Of the 148 million telephone calls attempted on the A.T.&T. network, about 56 percent, or 83 million calls, were completed [Sim90]."

Many businesses were disrupted, and some revenue was lost. Since Jan. 15 is a national holiday, with many businesses and other services closed, losses were minimized. The failure in the switching system is an acute example of how complexity, and complex systems, can acquire an unpredictable, almost life-like behavior [Ste91].

> "The innate peril of this computer complexity was highlighted by the [A.T.&T.] breakdown because the telephone company's switching network was [designed] with exactly this kind of breakdown in mind. Even though the system was created to prevent any single failure from incapacitating the nation's telephones, that is what happened.
>
> "The telephone company's system has been cited as evidence that extremely complex computer systems can be virtually flawless. Indeed, some computer and military experts believe that a space-based anti-missile system composed of thousands of independent sensors and computers could act reliably.
>
> "Late yesterday, A.T.&T. officials said they had traced the problem to a faulty program running on a computer that determines which path a long distance call takes. The faulty program, a new version of the switching software, sent a swarm of overload alarms to other computers in the network, causing widespread congestion.
>
> "The computer network should have responded by finding alternative routes for the calls; instead, about half the long distance calls ended in busy signals or recorded messages for nine hours Monday (Jan. 15) afternoon and evening.

---

[35] Unless noted, all quoted text for the A.T.&T. failure is stated in [JMa90]. Copyright © 1990 by the New York Times Company. Reprinted by permission.

"A.T.&T. has still not explained why the system failed to compensate for the failure of a single component, but the failure is seen as troubling evidence of how vulnerable technology-based systems are to even slight disruptions."

The problem source was eventually pinpointed, and corrected. The latent fault materialized, and the failure emerged after a sufficient volume of calls had passed through the system. After many days and months of continuous use without a failure of this magnitude, the proper superposition of conditions contributing to failure were met. To test for this possibility would have required the simulation of billions of telephone calls, and would have taken an enormous period of time to conduct.

So far as software safety is concerned, the ideal situation would be to preclude these conditions from ever arising. But a system composed of millions of lines of software, written by hundreds or thousands of individuals cannot be economically tested [Bea90]. Thus, some level of uncertainty, and a reliability figure less than one, will always exist in a system which is so large and complex, that it cannot be exhaustively tested under controlled conditions.

Regional Bell Operating Companies, June 26, 1991[36]

In a similar incident, the telephone switching systems operated by regional Bell operating companies malfunctioned first in Los Angeles, then Washington D.C., Virginia, Maryland, and West Virginia. On July 2, Pittsburgh and then San Francisco suffered similar failures. Each region relied on switching equipment manufactured by DSC Communications (Plano, TX).

"At the least, the cascade of disruptions indicates that the breakdowns are not coincidental. DSC Communications makes equipment for five of the seven Bell companies, and similar networks employing Signaling System 7 are made by the American Telephone and Telegraph Company and Northern Telecom.

"While engineers from Bell Atlantic and Pacific Telesis struggled to pinpoint the origins of their problems, other telephone companies went on alert to protect their own networks. Ameritech in Chicago, for example, said it had heightened monitoring efforts and would delay installing any new software until it had a full report from Bell Atlantic and Bell Communications Research, a consortium owned by the regional Bell companies.

"Nevertheless, telephone officials were clearly perplexed and frustrated as seemingly minor problems caused chaos throughout their networks.

"'Our software should deal with all these little maintenance occurrences,' said

---

[36] Unless noted, all quoted text for the Regional Bell Operating Companies failure is stated in [And91]. Copyright © 1991 by the New York Times Company. Reprinted by permission.

John Seazholtz, vice-president for technology at Bell Atlantic, which suffered a burned out circuit board. 'Why is our system getting overloaded everytime we get some rinky-dink little problem?'

"Although much of the evidence points to a software defect in the equipment supplied by DSC Communications, it was unknown why all the outages occurred in such a short time, given that the telephone companies have been using these systems for several years.

"The seemingly minor defects that caused a cascade of problems were different. That raises the specter of a more basic defect in the software that can be activated by a number of different things."

An analysis of the failure revealed that three lines of software were incorrectly coded. Engineers at Pacific Bell, the telecommunications subsidiary of Pacific Telesis, were able to reproduce the observed signalling failure, and wrote a corrective patch. That all the failures of the DSC signalling computers occurred within a short time frame actually resulted from a very recent but minor patch that had been installed. DSC unleashed the patched version into the network without running their standard 13-week acceptance test.

"The glitch was introduced when [DSC] made a minor modification in the program that runs its sophisticated call-routing computers, which are used by five of the seven regional Bell companies. In making what seemed to be an innocuous change, DSC dropped several algorithms, or processing instructions, that apparently caused the computers to go berserk when they experienced routine malfunctions."

"Officials at DSC admitted that they had not put the software upgrade through a customary 13-week test, because the change entailed only a few lines of code."

Each of the four software failures highlights an unintentional omission by the software engineering organization to implement correct quality procedures or design practices. These idiosyncracies amounted to a severe blow to corporate prestige, and in the case of the Therac 25, death.

The human element is clearly a big unknown in the software engineering equation. And since software plays an ever greater role in our society, from controlling anti-lock brake systems, a B-2 Stealth Bomber, automated teller machines and bank networks, and air-traffic control systems, one can reasonably expect that continued disruption in service, monetary losses, and even death will increasingly visit the population which is dependent on a computerized infrastructure.

Monetary losses and death pre-date the software revolution, but the acceleration and rise in software-related or dependent incidents will become more visible unless a mechanism for forestalling electronically induced collapse and catastrophe is found. It is for this reason that software safety has become a vital subject, and should not escape recognition by software practitioners.

5.2 Safety Categorization

Leveson [Lev83] enumerates the operational states of a system into 4 distinct categories. Each expresses an instantaneous measurement of safety and performance. The first state represents "correct and safe" operation, where the system is operationally nominal, the second is "correct and unsafe" where the system is operationally nominal, but an unsafe state threatens completion, the third is "incorrect and safe" where the system experiences operational disruption, but is not in an unsafe condition, and the fourth is where the system is malfunctioning and potentially capable of producing injurious outcome.

An example of the first state might be a train headed for a destination, on time, and within proper speed and fuel consumption parameters. For the second state, the train may encounter another headed in the opposite direction on the same track, but sufficiently far away that danger of a collision may be avoided. The third state may be the detour of the train onto another track to avoid the collision. The other track might temporarily displace the train from an ideal course, but a collision will be avoided.

These states constitute a means to describe system safety. They are a taxonomy of safety conditions in an operational environment. While 'correct' and 'safe' or their antonyms are not definitive mathematical terms, they convey the desirous properties to build into a system.

In each case, the term 'correct' implies operational or functional performance with respect to a specification or preordained directive. System correctness can be assured through a formal specification; a mathematical proof of operational attributes and functional behavior, provided that the specification is successful and correctly embodied in the underlying software representation. Formal methods are discussed in section 5.3.

5.2.1 Error Introduction and Safety Compromise

The four software failures discussed in section 5.1.2 illustrate several entry points that can enable a system to change from a safe to an unsafe state. In the Therac 25, both design and operational human error combined with a latent fault to generate an unsafe and incorrect system. The TCAS failure points out how design flaws propagate through an engineering activity and emerge as latent failures at operational stages. The TCAS incident could have been avoided if a test condition were added to simulate large flocks of air traffic.

The A.T.&T. failure in January, 1990, is still largely unexplained. If the system was designed as fail-soft, why did it paralyze the entire network? Clearly, the fail-soft mechanisms were inadequate. Perhaps the fail-soft procedures were not adequately tested, or their design was poor.

The local operating company failures of July, 1991, were traced to the introduction of an erroneously patched software module. No acceptance or regression testing was applied to the revision. The omission points to an organizational failure, similar in scope (though less catastrophic in terms of human life) to that which contributed to the Challenger space shuttle explosion in January, 1986.

Human fallibility is the locus of these failures. Techniques for minimizing the likelihood of unsafe system states are examined next.

5.3 Safety Analysis Methods and Requirements

Safety analysis has many goals. The most important is the detection of potentially unsafe or hazardous operating conditions which a system may enter. System safety assessment must be accompanied with notification and vociferous reporting to authorities or management. Speak up if you have verifiable evidence of unsafe conditions, or even suspect it. Failure to disseminate critical information related to system safety is a serious breach of ethical and moral conduct, especially when serious consequences may result from a system failure.

Since a system's operational safety characteristics embody substantial value, and serve as a beacon of quality, the process of isolating unsafe operating states, whether functionally correct or incorrect, must begin with system requirements and continue throughout the entire product lifecycle. From the United States Department of Defense standard DoD-Std-882B on System Safety Program Requirements [DoD84],[37] we find that system safety analysis should

> "identify the hazards of a system and impose design requirements and management controls to prevent mishaps by eliminating hazards or reducing the associated risk to a level acceptable to the managing activity. The term 'managing activity' usually refers to the Government procuring activity, but may include prime or associate contractors or subcontractors who wish to impose system safety tasks on their suppliers."[38]

DoD-Std-882B emphasizes the establishment of a safety assessment program which participates, monitors, and advises design engineering, manufacturing, and testing organizations about the operational concerns and professional practices applied to a product development process. System safety engineers, who require education and skill to apply professional scientific and engineering principles, criteria, and techniques to identify and eliminate hazards, or reduce the risk associated with hazards, must

---

[37] The British Ministry of Defence has a draft standard, DefStan00-55 (1989), directing the use of formal methods for safety-critical systems development. Ministry of Defence (UK), Critical Software Steering Group. Interim Defence Standard 00-55 (Draft), May 1989.

[38] DoD-882B uses the term 'hazard', which it defines as "a condition that is prerequisite to a mishap." Within the context of this chapter, a hazard is considered equivalent to either a fault or unsafe state. A 'mishap', defined by DoD-882B as "an unplanned event or series of events that results in death, injury, occupational illness, or damage to or loss of equipment or property" is equivalent to a failure.

exercise autonomous judgement in safety matters.

But system safety, design, and manufacturing are linked by the common thread of human fallibility. This bond implies that while certain unsafe requirements, design practices, manufacturing methods, or testing procedures will be immediate and obvious, latent and subtle faults may not be discovered until a failure arises. This circumstance and possibility were realized by all of the failures discussed earlier. Although many faults were likely detected during acceptance testing and corrected, some crept into the operational products. Undetected faults which enter into the operational software product and eventually lead to failure are representative of human engineered systems.

5.3.1 Requirements

DoD-Std-882B outlines the requirements for establishing a system safety program plan (SSPP). Many of these requirements pertain to system components other than software, such as material selection, fire or explosion, electromagnetic interference, and other possible sources of hazards and failures. The preliminary hazard analysis (PHA) is used to "identify safety-critical areas, evaluate hazards, and identify the safety design criteria to be used." [DoD84, pg. 202-1]

With respect to software, and software-dependent subsystem/system mishaps, the standard mandates that

> "safety design criteria to control safety-critical software commands and responses (e.g., inadvertent command, failure to command, untimely command or responses, or other operational characteristics) shall be identified and appropriate action taken to incorporate them in the software (and related hardware) specifications." [DoD84, pg. 202-1]

The PHA activity should therefore accompany a "concept phase"[39] product specification, and documents the projected outcome of failures due to software, hardware, material, etc. The PHA should serve as a guidepost for downstream development, engineering, and manufacturing processes. All findings in the PHA should be objectively satisfied throughout the system lifecycle following concept phase development.

Following the concept phase, a system specification is constructed. From this document, all pertinent subsystems will be identified, their roles defined, and their interfaces noted. A subsystem hazard assessment (SSHA) is conducted which identifies the failure modes for the independent subsystems, and failures due to interface faults. The SSHA is a rigorous process, and includes a determination

> "a. Of the modes of failure including reasonable human errors as well as single point failures, and the effects on safety when failures occur in subsystem

---

[39] The term 'concept phase' implies exploratory design or development of a system's operational representation through a computer program simulation or scale-model mock up. It does not imply that a miniature nuclear reactor be constructed.

components.

"b. Of the potential contribution of software events, faults, and occurrences (such as improper timing) on the safety of the subsystem.

"c. That the safety design criteria in the software specification(s) have been satisfied.

"d. That the method of implementation of software design requirements and corrective actions has not impaired or decreased the safety of the subsystem nor has introduced any new hazards." [DoD84, pg. 203-1]

The SSHA points out the necessity to isolate and identify sources of human error which can lead to failure. The organization which designed the Therac 25 dosage editor did not heed this requirement. It is doubtful that they read and understood DoD-Std-882B.

Both telephone network failures illustrate how 'corrective actions' were inadequate and not robust. Corrective actions implies exception handling, such as that provided for by some programming languages and OSs. Exception handling is useful when undesired outcomes or results are generated *in medias res*, and alternative steps must be enacted to prevent a complete system breakdown.

Leveson [Lev81] has suggested several possible mechanisms for implementing corrective actions to ensure fail-soft execution. Among these are a "safety kernel," which is akin to an OS kernel implementing security features. The safety kernel may be empowered to attempt recovery of critical functions, or activate reconfiguration operations on damaged modules.

Leveson [Lev81, pg. 16] also suggests that critical functions be isolated from non-critical ones, and that a "supervisor" possess the autonomy to reconfigure the system. Some errors, such as floating-point overflow or a bad computation may be non-critical, and not affect system safety. But others, such as an infinite loop, may drastically alter operational effectiveness (by causing guidance failure). Can a supervisor suspend or cancel non-critical functions, and improve operational safety? This question has yet to be resolved in practice.

For software system safety, DoD-Std-882B requires the analysis of software at seven different stages of design, development, testing, and maintenance. The software requirements hazard analysis (SRHA), top-level design hazard analysis (TDHA), detail design hazard analysis (DDHA), code-level software hazard analysis (CSHA), software safety testing, software/user interface analysis, and software change hazard analysis. Each of these tasks is used to assure software safety through inspection, walk-throughs, analytical techniques, and testing. Detailed explanations for all stages are located in the standard.

The standard recommends several quantitative analysis techniques and methods to assess software safety. Some of these methods for software safety analysis are discussed next.

5.3.2 Standard Methods for Software Safety Analysis

Analysis tools for software safety purposes stress the detection of unsafe states. Several methods are known. DoD-Std-882B explicitly mentions software fault trees and Petri nets as viable alternatives for assessing software safety. Each method discussed here has specific strengths and weaknesses. For greatest effectiveness, more than one should be used for analysis purposes. Redundant analyses serve as a sanity check, and can uncover more unsafe states.

The process of software analysis is illustrated in Figure 5.1. The analysis of a system begins with its specification of function and purpose. This specification is abstracted or cast into a modeler which realizes the abstraction as an executable model which embodies the design and intended functional properties. The system simulator executes this model, generating results. The simulation results are then checked against the desired behavior, which is represented by a separate set of expected outputs, conditions, or formal specifications. These two sets of outputs are compared during the results analysis phase [Gua90].

Checking the system's simulation output versus that specified by a formal specification or acceptance test criteria reveals the correspondent behavior between the implementation as created by an organization and the functional objectives of quality, reliability, and safety.

The software analysis process as currently practiced is largely a manual one. Few, if any, automated techniques for software exist. Hardware analysis is different, in that fault detection can often be found through simulation via VHDL-based specifications. The software fault tree analysis method presented here is based on the notion of hardware fault tree analysis. An automated tool which operates on specifications and simultaneously assesses run-time simulation results to produce fault trees and safety analysis would be a tremendous asset.

However, no guarantee can be placed on the absolute correctness of input specifications.

> "Formal proof techniques are an attempt to remedy some of the limitations of testing. Even assuming that mechanical verifications were available and usable and that cost was unimportant, a formal proof only demonstrates the correctness of the program with respect to the specifications. There is no guarantee that the specifications are correct, and in fact, writing formal specifications is a difficult and error-prone process [Ger76]. Furthermore, the complexity of proving a large system correct is so great that the process must itself be error-prone." [Lev81]

Despite the uncertainty present in the software engineering process due to the intangible human element, analysis techniques can be successful, even when conducted without automated assistance. Leveson's remarks on formal specification are pointed and accurate. Formal specification practices have largely focused on critical applications, such as a nuclear reactor, or military systems. These edifices certainly qualify as complex.

But the stigma associated with formal methods is diminishing. Formal methods are finding their way into commercial applications. Further discussion on formal methods is deferred to section 5.4.

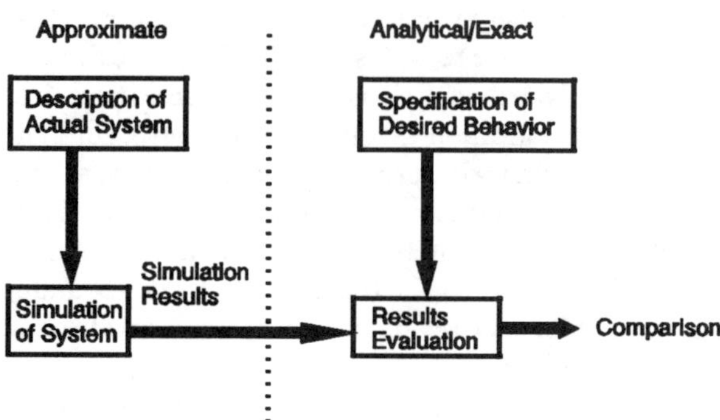

Figure 5.1 Software analysis process (after [Gua90]).

5.3.2.1 Petri Net Analysis

Petri nets are useful for modelling systems. A Petri net representation of a system becomes an abstract mathematical entity which simulates its behavior and general functional properties. From Peterson [Pet81], "Analysis of the Petri net can then, hopefully, reveal important information about the structure and dynamic behavior of the modeled system." Like all simulations, a realistic approximation of the system is sought.

With respect to software safety for real-time systems, Leveson [Lev87] used a variant of Petri nets -- Time Petri nets -- to analyze safety and fault tolerance by constructing a network, and then subjecting the network to an algorithm which identifies and eliminates critical states from a list of previously known "high-risk conditions." A *reachability tree* is produced which shows the paths that a system can follow toward unsafe conditions [Lev87]. Unsafe system states can be extracted from reachability trees, and this information can be used to design safeguards and fail-safe mechanisms into the system, and prevent serious consequences from arising during a failure.

Constructing reachability trees is an advanced topic in Petri net theory, and an essential process for identifying unsafe system states. Tree construction is thoroughly discussed by Peterson [Pet81, Chapter 4].

Detecting how a system can evolve into an unsafe state is a vitally important design goal, especially for very complex installations, such as power plants and refineries. Analyses of Petri net representations can provide this information. Only a rudimentary discussion of Petri net characteristics, properties, and usage is given here to acquaint the reader with this analysis technique.[40] A good introduction to Petri nets for timing requirements assessment is that of Coolahan and Roussopoulos [Coo83].

A Petri net is composed of five entities: a set of *places* P, a set of *transitions* T, an *input* function I, an *output* function O, and an *initial marking* μ_0, and is denoted as C= $\{P, T, I, O, \mu_0\}$. *Tokens* define how the net executes. Tokens, a primitive concept in Petri nets, are restricted in movement between transitions according to the input and output functions. Transitions can fire only when there are at least as many tokens at the input places as there are arcs from a place to a transition.

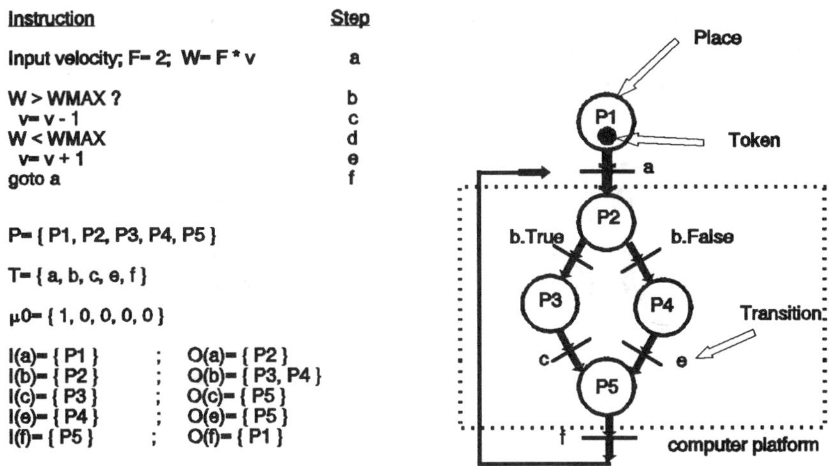

Figure 5.2 A simple Petri net graph with initial marking μ_0, places, input, output, and transitions.

The input and output functions map transitions between places of the net. The linkage between transitions defines how a computation is conducted. Tokens visit specific places along by traversing between input and output function mappings; the tokens embody the mathematical representation of the system and the computational conducted within. The initial marking, μ_0, defines the location of tokens before the net fires; μ_0 represents the initial conditions of the system. Figure 5.2 illustrates a simple

[40] The reader is referred to [Pet81] for a thorough presentation of Petri nets and Petri net theory.

Petri net.

The figure shows how a small code fragment for a control limit check is cast into a Petri net. The limit check is designed to keep the power level constant, assuming a constant application of force, and does so by adjusting the velocity variable in either step C or step E. The token circulates through either the TRUE or the FALSE path represented with transitions b.TRUE or b.FALSE ([Pet81, pp. 57-58]) This decision is carried out in software.

The computer platform is assumed to be separate from the force and velocity sensors. This is typical in many situations, where sensor output is directed through an I/O port into the computer, and the computer operates on the data.

In another case study, Nelson *et al.* [Nel83] engineered a programming system for generating executable code from a Petri net specification language. This system produced PL/I and PL/S code which was compiled into a program on a mainframe computer. The motivation for this study was to demonstrate how a Petri net representation of a concurrent, multiprocess system could be translated into an executable target for a multiprocessor platform.

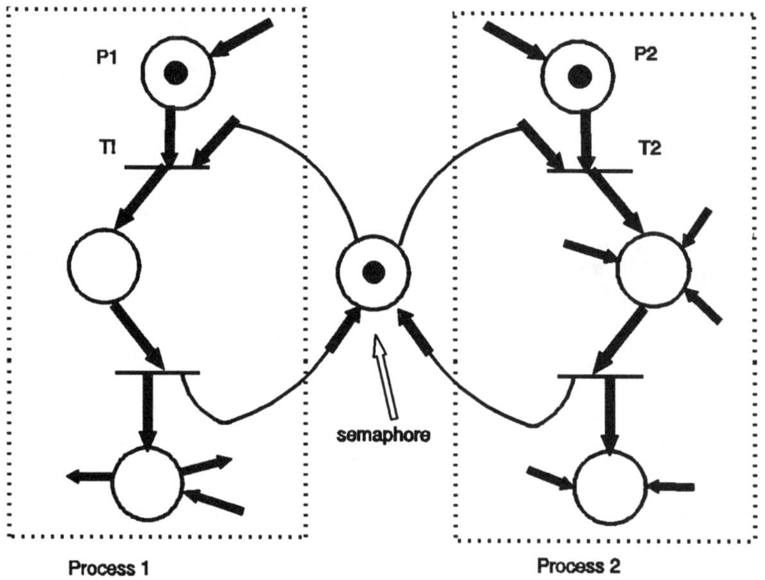

Figure 5.3 A Petri net representation of two processes mediated by a semaphore.

Petri nets offer an expedient mechanism to represent concurrent process structures. This property is illustrated in Figure 5.3 ([Pet81, pg. 63]), where two processes use a semaphore to mediate contexts and synchronize execution.

The synchronization point, as illustrated here, is controlled by a semaphore to protect each process during access of a shared object. The *critical section* ([Pet81, pg. 62]) of each process must be protected from interference by the companion; otherwise, specious results may be generated by the computation. The semaphore is an example of a locking mechanism implementing *mutual exclusion* ([Pet81, pg. 62]).

Figure 5.3 shows how the semaphore enforces mutual exclusion. For the Petri net to fire, two tokens must be located in either process simultaneously. The places P1 and P2 each have one token. Either net will acquire both tokens, when the semaphore transfers an "enabling" token through the transitions T1 or T2, leaving the companion in a blocked state with only one token.

One process will then own both tokens, which enables the net to fire, and the computation represented by the other unnamed places will be carried out downstream of the critical section. The semaphore token is returned to its "home" location, and then becomes available for the other process to execute without interference.

This simplified example is a very modest formulation. Petri nets for large systems can be extraordinarily large, and difficult or nearly impossible to evaluate. Complex systems can acquire an opacity which prohibits an analysis of a Petri net mathematical representation. Petri net modeling is not an easy technique for the novice to apply. Determination of reachability criterion, and reachability tree construction are methods beyond the scope of this text. They require considerable expertise, practice, and patience to correctly exercise.[41]

5.3.2.2 Software Fault Tree Analysis

As a method for software safety analysis, software fault tree analysis (SFTA) is best applied to the design or code phases of a system lifecycle. Leveson [Lev83] states, "The goal of SFTA is to show that the logic contained in the software design will not produce system safety failures, and to determine environmental conditions which could lead to the software causing a safety failure." Fault trees are simple to construct, and can be automatically derived from high-level design or code specifications, such as a program design language (PDL).

Using a PDL as a design specification instrument provides a mechanism to organize hierarchy within the design, and thus develop greater levels of abstraction which can be subjected to SFTA. Abstraction is tightly coupled to the implemented code, so SFTA applied directly to code embellishes the results with a blend of detailed logical failure paths at the statement level to failure modes affecting system components.

Figure 5.4 illustrates a fault tree for an if-then-else statement with attendant source code. The tree is constructed from symbols found in Leveson [Lev83]. Each branch of the if-then-else statement is cast as input to an AND operator. One input to the AND for each branch is needed for each conditional outcome. The resultant output produced from the if-then-else branch specific statement is input to an OR operator.

Fault trees constructed from AND/OR functions are termed *coherent*; trees including NOT or XOR (exclusive OR) functions are called *non-coherent* [Chu90].

The left branch of the tree, which corresponds to the condition "speed ≤ MAXSPEED" prior to the conditional contains an unrealizable input. Under the

---

[41] More than this author was willing to invest for this chapter.

assumption that excess speed produces a failure, this fault tree describes the "loss event."[42]

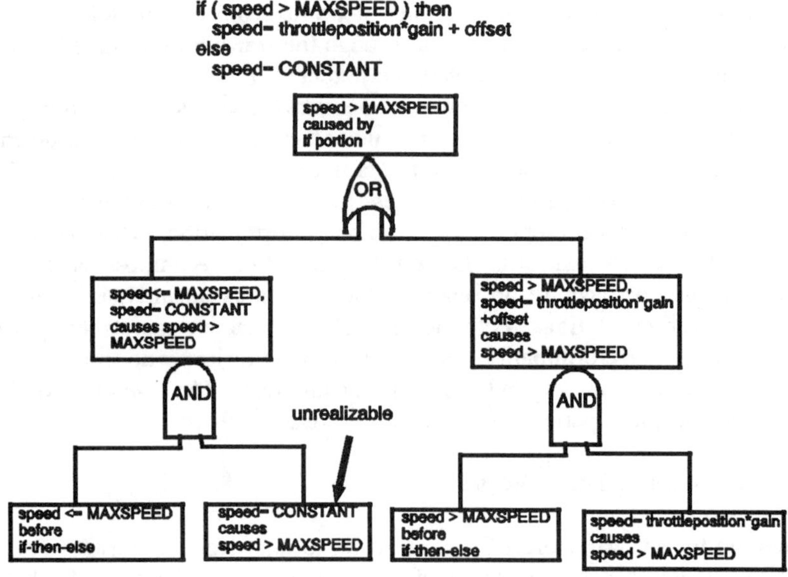

Figure 5.4 Fault tree for an if-then-else construct.

A loss event can be pinpointed via SFTA, and the failure isolation characteristic provides an attractive argument for employing this method during hazard analysis [Cha88].

At the code level, SFTA provides a schematic of faults, as illustrated by the fault tree itself. However, the conditions contributing to potential loss events or failures, such as input parameters and timing constraints are not automatically assessed. How the system behaves under varying input conditions is not immediately identifiable.

5.3.2.3 Logic Flowgraph Methodology

The logic flowgraph methodology (LFM) is a recent development ([Gua88], [Gua90], [Gua91]) which is often used in combination with SFTA. LFM is the centerpiece of automated decision support systems used to control complex installations, like nuclear power plants. Guarro *et al.* [Gua90] explains that

> "the application of LFM is typically a two-stage process. The first stage is the construction of a model for the system of interest. This model is derived by

---

42 [Lev83], pg. 571.

> identifying first all the basic process parameters by which the system behavior can best be described and then by expressing the fundamental cause-effect interactions (and the conditioning effects on these interactions effected by faults and operational mode changes) among those parameters. The second stage is the search for the combinations of events and conditions which may produce certain process states (identified by the values that certain parameters may assume within the given process). This second stage is executable in the form of an automated procedure which traces cause-effect relationships in reverse through the LFM network model."

Building a model of the system description which LFM can process and reduce to an executable context speeds application of this technique. Currently, LFM relies on the systems engineer to construct a graphical representation through a CASE tool. While structured methods are standard practice in DoD environments, CASE tools often provide a multi-orthogonal representation of the system. Entity relationship [Che89] and dataflow diagrams ([You89] and [Gan89]), and data dictionaries often appear side by side as equivalent, but often equally inexact representations of the system.

This dilemma forces the system designer to implement an ad hoc simulation of the system, implemented through a conventional programming language, to represent the principal elements of functionality to examine via LFM. But once the system has an adequate representation, LFM analysis produces a

> "documented model of system behavior and interactions, which fault tree analysis does not provide nor document directly. From the LFM system model it is possible to deduce, by means of the automated back-tracking procedure, a comprehensive set of fault trees that may be of interest for a given system. This is a rather notable feature, since, once an LFM model has been developed, it is not necessary to perform separate model constructions for each top event of interest (as is instead the case in the application of fault tree analysis). Because LFM modeling focuses on parameters, rather than 'components', it also offers greater modeling flexibility than fault tree analysis, although this flexibility goes along with a proportionally greater complexity of the modeling rules and syntax with respect to fault trees." [Gua90, pp. 25-26]

LFM combines with SFTA for an important purpose: the LFM model developed by the system designer treats both hardware and software components identically, and as a coherent unit rather than independent and decoupled entities. With a unified representation of both hardware and software system components, the LFM model is cast into directed graph representations, where causal or sequential interactions among parameters can be evaluated and searched to reveal high-level hazardous system states, represented by fault trees, involving hardware and software components.

Identifying system parameters that correspond to external physical measurements or variables and/or interfaces between the software and the hardware components is a central concern for LFM model construction. Parameters such as engine thrust and fuel flow or combustion temperature are pivotal to correct jet engine function, and an LFM model of such a system would certainly utilize and incorporate these parameters; they form the vertices of the directed graph. A high level control flow diagram (CFD) is a visual by-product of LFM graph construction.

Once the high-level hazards have been identified, the software dependency for a hazardous state can be investigated in greater detail via SFTA. For SFTA to operate within the LFM framework, further refinement of the high level control flow diagram is needed. A local control flow diagram (LCFD) is created for each subroutine. The LCFD expresses conditional branches, loops, and other control structures.

The LCFD is used to perform a control path search (CPS) to identify which parameters in the software possibly contribute to the hazardous states produced by the LFM fault tree constructed from the CFD. "Through each control path, the relationship between input parameters and output parameters of a particular procedure can next be identified, which accomplishes the relationship search portion of the LFM-SFTA approach. Control path and relationship search procedures are continued through the software, until basic hardware inputs are reached." [Gua90, pg. 32].

The UCLA Embedded System Software Assessment and Enhancement (ESSAE) [Gua91] Project is spearheading the investigation into LFM-SFTA to "identify and develop modeling techniques which can be used to verify, in terms of system reliability and/or safety, the correspondence of the desired behavior of an embedded system with the behavior which can be expected once an actual design of the system has been developed and implemented." [Gua90]

If LFM-SFTA can be organized into an effective product, investigation of software safety will become greatly automated, provided that a unified methodology for LFM system specification accompanies the resulting analysis tool.

5.4 Formal Methods and Real-time Software Safety Issues

A *formal method of specification* encompasses the process of constructing an unambiguous functional description of a system by using mathematical principles and proof. The language of mathematics is non-contextual; semantics do not influence the interpretation of the stated derivation, theorem, lemma, or definition. A *formal specification* consists of mathematical statements, usually stated in the form of logic and first-order predicate calculus, along with a descriptive translation of the mathematical logic.

> "Formal methods are used to reveal ambiguity, incompleteness, and inconsistency in a system. When used early in the system development process, they can reveal design flaws that otherwise might be discovered only during costly testing phases. When used later, they can help determine the correctness of a system implementation and the equivalence of different implementations." [Win90]

The Therac 25 incident resulted from a poorly designed dosage editor. Did the Therac 25 software specification formally detail how to prevent or detect corrupted operator input? Apparently not. And so this fault was propagated through the design cycle into the implementation. Formal specification practices seek to prevent accidents like that with the Therac 25 from ever occurring.

But formal specifications are not computer programs, and they are created by humans. The most ideal situation would have formal specifications undergo *transliteration*, translation from the specification language into the implementation software for compilation at the push of the button. Automatic programming is an elusive goal, although some approaches have been researched [Fre87]; almost all are based on specification languages, but not formal methods of specification.

Formal methods of specification have been used in several instances. The Inmos transputer instruction set and floating point unit have been subjected to formal methods of specification and verification by simulation prior to the casting of masks for production of silicon implementations [INM88]. Although post hoc, the UNIX filing system has been formally specified [Mor84]. A real-time kernel has been defined via Z and formal methods [Spi90]. Numerous other examples can be found in Hayes [Hay87] and [Nic90].

Stankovic [Sta88, pg. 16] has stated that "the fundamental challenge in the specification and verification of real-time systems is how to incorporate the time metric. Methods must be devised for including timing constraints in specifications and for establishing that a system satisfies such a specification." In 1986, the author participated in the construction of a real-time simulation where the scheduling parameters, frame-times, and dispatching intervals were decided by committee; this approach satisfied the customer, who was merely concerned with fidelity and realism. Our goal was to achieve an implementation which realized fidelity, without consuming too much machine resource.

Our fidelity requirements were satisfied with a great deal of tuning, tweaking, and overall sleight-of-hand. Today, constructing a specification which not only formally asserts algorithmic or computational correctness, but includes and articulates temporal performance is indeed a 'challenge.' One specification method, the RT-ASLAN language and specification process developed by Auernheimer and Kemmerer [Aue86], incorporates abstraction techniques and language elements which permit the declaration of deadline constrained functions and processes. The RT-ASLAN system can generate "performance correctness conjectures"([Aue86, pg. 879]) as well as functional correctness conjectures, which are conducted through an induction process involving the initial state of the specification.

RT-ASLAN is not based on a formal mathematical specification framework, but instead uses a specification language based on a simple grammar, and is more like a process algebra, where a model is principally expressed as a system of processes which communicate with each other [Hoa85]. The grammar is used to abstract a system, components, and timing requirements into a specification which is then analyzed for timing inconsistencies and functional aberrations. The effectiveness of RT-ASLAN in the design of a realistic system has yet to be verified.

Formal methods are applicable to more than the systems requirements definition of a lifecycle. During detailed design, where algorithmic development is needed, formal definition of the implementation can save resources during testing. Software engineering is an expensive process, and any method which can eliminate costs and provide improved product quality should be employed [Swa91]. But cost justification is a primary concern of department managers, and unless the payoff is clear, motivation for undertaking the investment in formal methods may be absent [Cop90].

Formal specifications can be used to implicate a contractor's failure to deliver agreed upon capabilities and functionality. Professional liability is a reality, as the U.S. case of Diversified Graphics vs. Groves [Blo89] indicates. "Formal specifications provide a factual basis for litigation, by defining the developer's responsibilities clearly and provably [Swa91]."

Although relatively few systems have employed formal methods of specification, the technique is increasingly popular, especially in the U.K. The Programming Research Group (PRG) at Oxford University has spearheaded much of the research into formal methods using the Z language [Spi88]. The PRG continues to lead much of the world in formal method research, and has done exceptionally well at promulgating this facet of software engineering discipline.

5.4.1 The Z Specification Language

Z[43] is a specification language based on typed set theory and first-order predicate calculus, together with a structuring method called the schema calculus. Z is an ideal vehicle for formal specification construction, although some education is required prior to undertaking such an effort. Object-oriented extensions of Z have been developed to accommodate information hiding, polymorphism [Str86], and class definitions to the hierarchical specification methodology of Z [Duk91]. The Z (Object-Z) schema is the principal structure for specification development and statement.

Adding temporal logic -- time-dependent operators -- to a Z specification has been alluded to by several principal researchers in formal methods [Nar90]. Auxiliary variables, entities which can denote historical change (in either the past or future sense), provide a temporal dependence to a Z specification. The concept of time is not explicit within Z. Constructing an Object-Z class for an interval timer is an alternative mechanism for implementing temporal dependence. Event histories can be constructed, and predicate operators can be used to restrict their ranges or limit their extent between specific time intervals. Until a formal Z construct for temporal logic is available, all approaches to implementing concurrent systems will be unique.

Concluding Remarks

The construction of safe software requires strict discipline towards the application of techniques for assessing hazardous conditions that can arise during system operation. Unfortunately, the availability of general purpose tools that can be purchased off-the-shelf to assist in safety analysis is largely non-existent. This implies that manual methods and tools must be used in their place. Recognizing software safety as a principal and visible component of design, implementation, and testing can reduce the likelihood of catastrophic failure. Human-engineered systems are likely to continually

---

[43] First introduced by Jean-Raymond Abrial [Hay87].

suffer from the foibles introduced through an evolutionary process which adduces change and advancement while sacrificing functional and absolute correctness. Face it, our reasoning processes are not perfect. So then, the systems we construct will not be perfect either. It is unfortunate that fellow beings will undoubtedly suffer from our collective ineptitude and imperfections.

Suggested Reading

John D. Musa [Mus89] provides an excellent introduction to software reliability estimation and modeling. Peterson [Pet81] gives an excellent treatment of Petri nets. An excellent introduction to formal specification is found in Potter, Sinclair, and Till [Pot91]. The September, 1990 issue of *IEEE Software* contains several case studies of formal specifications, and advice on how to integrate this technique in the software development process [Kem90]. The IEEE has also produced an excellent document that gives an encyclopedic list of techniques for assessing reliable software. IEEE Standard 982.2-1988 [IEEE88] describes reliability measures, and can be used as reference guide for software reliability assessment. The recent book by Leonard Lee [Lee91] catalogues many instances of software failure. See also [Lee92].

For those readers with USENET access, the newsgroups comp.specification and comp.specification.z contain postings concerning safety and formal specification practices. In Europe, the United Kingdom is spearheading the software safety movement, with the Oxford Programming Research Group in the lead. Bowen and Stavridou [Bow92] have compiled an excellent bibliography on software safety, formal methods and standards.

Part II

Multicomputer Methods

6

Multicomputer Software Design Issues

In this chapter, a technique for designing multicomputer software is presented. The basis of our multicomputer software design model resides with the notion of communicating sequential processes (CSP) developed by C.A.R. Hoare [Hoa85]. The author's experience in this domain arises from a fortuitous encounter with Occam, the Occam model of programming, and the transputer. Every attempt to divorce the discussion from a particular brand of hardware has been made. Our discussion relies on the properties of logical concurrency, as embodied by CSP. Logical concurrency abstraction is the principle approach used in this chapter for multicomputer software design. The goal is to construct host-independent simulations. Multicomputer hosts are widely available, and this commodity-like status lessens their role in the software engineering equation. Software imbues value in a computer system, and is especially important in the multicomputer case.[44]

The design technique is based on the analysis of process structure graphs and the logical concurrency they represent. Message-passing is required to create effective multicomputer simulations, and this method of interprocess communication is also discussed. Techniques for building data and control parallel simulations on multicomputers are outlined. A brief discussion of deadlock, how it arises, and steps to avoid it are discussed. Logically concurrent process descriptions are transformed to physically concurrent contexts by decomposing the problem (e.g., algorithm + data) among the multicomputer nodes. A simple means of emulating the physical concurrency is outlined as a method for easing this transformation and simplifying the debugging process. The methodology espoused in this chapter relies on the notion of soft-channel emulation via a stand-alone router support library that is linked into the multicomputer simulation. A software example based on this method, for a simple two-dimensional domain, illustrates the technique.

6.1 Comparison of Sequential and Concurrent Software Engineering

A multicomputer simulation, like a sequential implementation, is designed and constructed with the aid of familiar software engineering techniques, methods, tools,

---

[44] It is often remarked that software without hardware is an idea, while hardware without software is a spaceheater.

and languages. But the subject of multicomputer software engineering has received much negative attention in the popular press. Press reports often cite the extraordinary complexity of building a multicomputer simulation. Anyone who has tried to build multicomputer software knows that the initial attempts are frustrating and consuming. But these experiences inculcate essential discipline, much like solving simple integral equations for homework exercises prepares a student to tackle more difficult systems. To successfully construct and execute a multicomputer software engineering effort, the engineering team must commit themselves with the dedication and desire to achieve, and they must also possess strong design discipline. One cannot shotgun a multicomputer simulation; they must be finessed, groomed, and sculpted.

The three ds (dedication, desire, discipline) -- also known as D^3 -- distinguish and separate multicomputer software engineering from the activities normally practiced by those who perform shared-memory software engineering. D^3 is needed for the shared-memory environment at diminished intensity. The shared-memory environment is a comfortable medium in which to develop and debug software: the entire machine can be sequentially stepped, global memory addresses are easily examined, communication mechanisms and harnesses are virtually transparent (e.g., shared-memory, messages, queues, sockets and pipes), I/O bandwidth is plentiful, and a single clock times and controls the instruction stream and process sequencing. Figure 6.1 shows a typical shared-memory multiprocessor schematic.

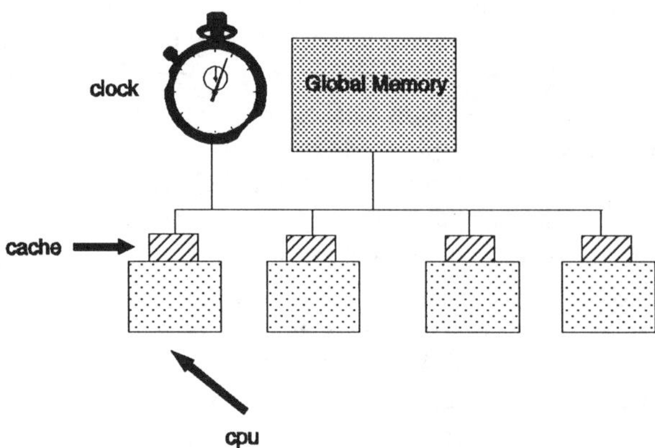

Figure 6.1 A shared-memory multiprocessor architecture schematic.

The single clock serves as the global time reference for all processing within the platform. Any process that inquires about time receives the value returned from this solitary clock source. Real-time simulations executing on shared-memory systems are usually not concerned with the reconciliation of time between multiple platform timers. Time is a global commodity in the shared-memory multiprocessor.

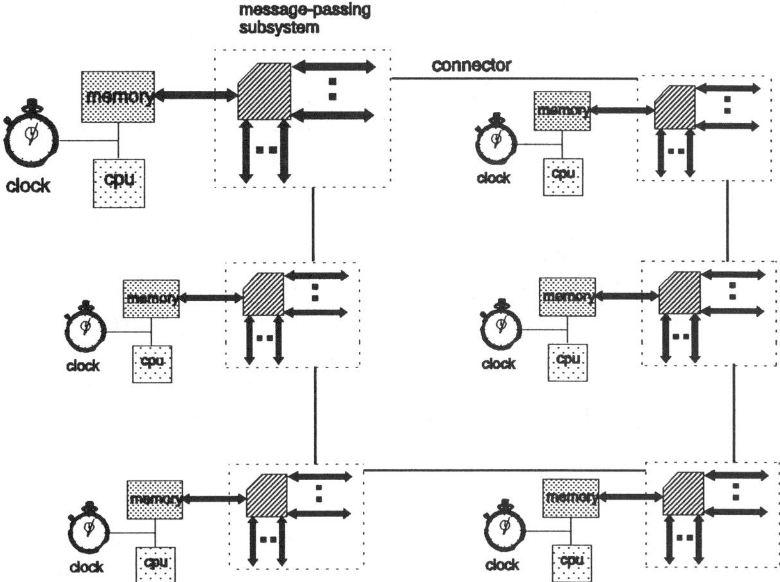

Figure 6.2 A toroidally connected multicomputer topology.

The multicomputer system illustrated in Figure 6.2 presents substantially greater challenges with respect to clock synchronization. For each *node*, a separate clock, memory space, processor, and message-passing subsystem is found. The clock in each node controls the basic memory access and instruction sequence functions found in all von Neumann-style computer systems, and also serves as the local time reference for that node. Time reference localization in the multicomputer creates a burdensome, but not impossible problem for developing real-time multicomputer simulations. The real-time multicomputer simulation must maintain a common global time reference to achieve predictable results. Processes referencing dissimilar time stamps will violate the predictability criterion discussed in Chapter 5. Failure to satisfy this criterion may cause deadlock, or lead to a software safety mishap.

The distributed memory structure of the multicomputer also forces the engineer to apply a different approach to algorithm and simulation design. While the shared-memory environment permits arbitrary placement of processes anywhere within the machine's address space, multicomputer implementations controvert this cavalier notion. Processes must be carefully placed for purposes of interprocess communication and message-passing efficiency.[45]

The principle difficulties to surmount when undertaking a multicomputer simulation are: the incorporation of message-passing primitives into the process structure, and the construction of a data decomposition which achieves a run-time load balance (see Chapter 7 on Load Balancing). These difficulties cannot be overcome without the aid of debugging techniques and tools. And in the case of real-time multicomputer simulation, the coordination and maintenance of a global time reference common to all local clock sources is a substantial problem. Clock synchronization is a necessary addition for real-time multicomputer simulations. A thorough discussion of clock synchronization is deferred until Chapter 8.

Training a seasoned software engineer to execute a multicomputer software engineering processes is not difficult. The first and most important hurdle to overcome in the re-education[46] process is to inculcate a concurrent thought process. Multicomputer contexts possess highly articulated process structures; many processes and communication activities can simultaneously execute during the course of a multicomputer simulation. Therefore, acquiring the means to conceptualize simultaneous processes and their interactions is essential to the multicomputer software engineering process. It is at this point that indoctrination begins.

6.2 Logical Concurrency[47]

A multicomputer simulation is a superposition of processes and communication primitives.[48] A process performs algorithmic work while communication primitives convey intermediate results and computed values to other process peers. A *process peer* is defined in terms of context pairs: processes that are assigned or acquire equal status within a hierarchy. The hierarchy of processes is most easily expressed in a structure graph. The *structure graph* (process structure graph), in the most crude formulation and level of detail, illustrates the process hierarchy and communication interfaces between each process peer. A process structure graph for a Laplace equation solver in two-dimensions is shown in Figure 6.3.

---

[45] Processes that share data and exist on separate processors are often placed on neighboring processors to minimize message-passing latency. With the arrival of packet-switched message-passing systems, latency is substantially reduced and process placement is less critical.

[46] The author wryly implies brainwashing.

[47] Critical among the many skills required of concurrent software engineers is the interpretation of the logical concurrency expressed by a process structure graph. This skill is like phrenology (the characterization of brain functions by examining the bumps on the cranium): the more you examine and design, the more skill you acquire.

[48] Communicating sequential processes (CSP). The Occam model of programming is CSP incarnate.

This graph states that four processes cooperatively execute a finite-difference computation over a two-dimensional domain. The two processes containing expressions of $U_{i,j}$ are algorithmically equivalent process peers of the same rank; they both compute partial derivatives and require the identical number and types of arithmetic operations to perform, only the data is different. The loop generator process and the derivative addition plus assignment process for $U^{new}{}_{i,j}$ share the same rank as the partial derivative computation processes. All processes enclosed within the graph (as circumscribed by the largest bubble) that appear at the same level are peers. The outermost bubble is the operating system and machine context.

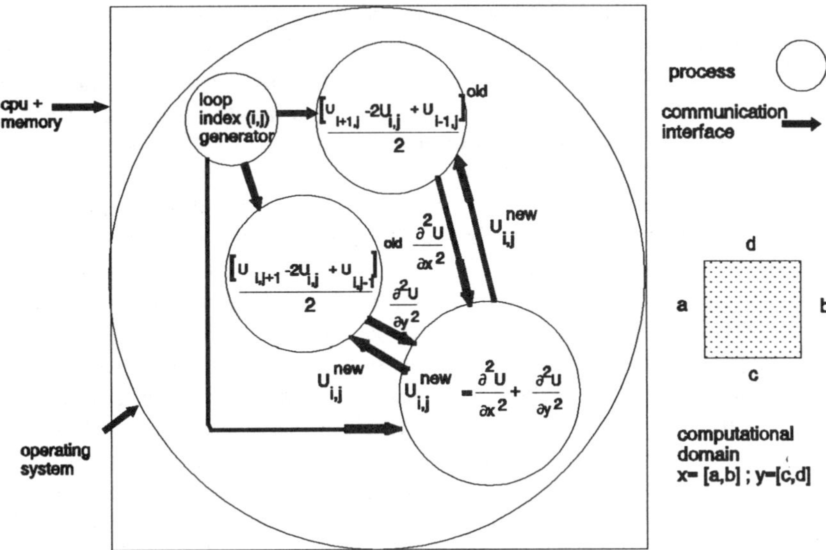

Figure 6.3 Process structure for a finite-difference solution of Laplace's equation using a two-dimensional computational molecule.

The process structure graph in Figure 6.3 graphically expresses an algorithm and the communication interfaces between the composite processes. The structure graph presents a logical simulation structure mirroring the desired physical simulation. Figure 6.3 declares that the loop generator process, the derivative addition-plus-assignment process for $U^{new}{}_{i,j}$, and the two partial derivative computations are scheduled to execute at the same time.

The communication pathway between the peers is filled by several *channels*. A channel permits the unidirectional flow of data from one peer, the sender, to a receiving process. Channels are memory-mapped structures shared between and among processes. Process peers are required to complete channel communications, or a deadlock situation will result (see section 6.4).

The process structure graph declares the organization of four concurrent processes, but they should not necessarily be implemented as a composite of four separate processes. That all four process peers share a common input (the loop index generator values i and j) means that each process is locked -- literally slaved -- to the loop index generator output. No work will be carried out by this process structure until the loop index generator process presents each process peer with an input. The computation is better suited to execute in a purely sequential manner.

The loop index generator drives the computation by serving the second-order partial derivative addition-plus-assignment processes with indices. Each pair of indices passed to the derivative algorithm generates either $\partial^2U/\partial x^2$ or $\partial^2U/\partial y^2$. At the same time, the derivative addition-plus-assignment process establishes the quantity $U^{new}_{i,j}$, and then blocks -- suspends -- until its input is satisfied. Each iteration of the algorithm will not go forward until the derivative computations finish, and the derivative addition-plus-assignment function receives the derivative process output values. The derivative addition-plus-assignment process blocks until the values are delivered along with the loop indices, where $U^{new}_{i,j}$ is finally concluded. Each phase of the computation for each loop index pair completes when $U^{new}_{i,j}$ is delivered to the derivative computation for the next iteration.

The *logical concurrency* expressed by a structure graph is merely the sum of the communication and process attributes. The logical concurrency order provides an heuristic metric (a rule of thumb) for organizing a match between the simulation and the underlying platform. The graph in Figure 6.3 describes a process structure with logical concurrency of order four for processes and 7 for communication.

All processes in Figure 6.3 are controlled by the generation of loop indices. This condition effectively reduces -- collapses -- the logical concurrency to order 1 for the entire graph. The reduction yields a single process like that illustrated in the pseudo-code -- a simple iterative calculation. The communication channels leading to and from the processes vanish; sequential access to the variables through traditional memory address patterns replaces communication channels. The collapse of logical concurrency from an order greater than unity to a value of unity implies that the computation may be replicated.

The computation is concisely stated by the following pseudocode representation (Equation 6.1):

for i=1 *to* n - 1 */* i=0 & i=n are boundary values */*
  for j=1 *to* m - 1 */* j=0 & j=m are boundary values */*
  {

$$\frac{\partial^2 U}{\partial x^2} = \frac{u_{i+1,j} - 2u_{i,j} + u_{i-1,j}}{2}$$

$$\frac{\partial^2 U}{\partial y^2} = \frac{u_{i,j+1} - 2u_{i,j} + u_{i,j-1}}{2} \qquad 6.1$$

$$U^{new}_{i,j} = \frac{\partial^2 U}{\partial x^2} + \frac{\partial^2 U}{\partial x^2}$$

  }

A unit concurrency measure corresponds to a sequential process structure, as the pseudo-code example shows. A simple double loop replaces the multiple processes and communications expressed in Figure 6.3. No clear advantage is gained by implementing the order four process structure owing to the synchronizing authority imposed by the loop index generator over the entire graph. The collapse of logical concurrency illustrated here supplies an important opportunity to redesign the Laplace equation simulation to achieve greater performance through process structure replication.

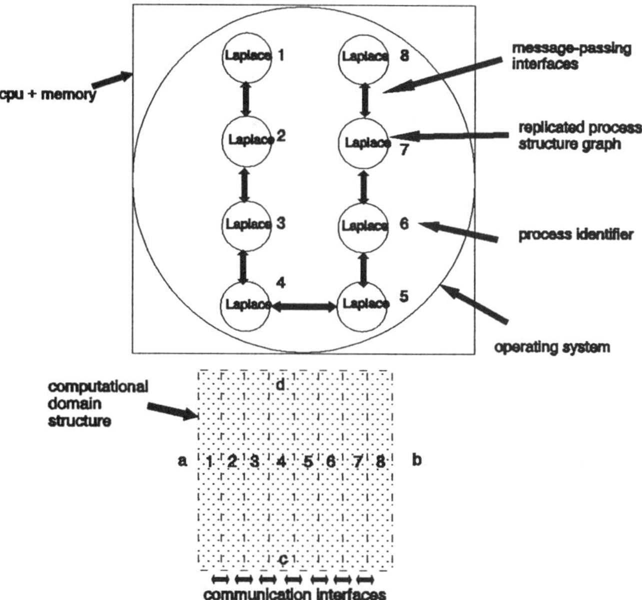

Figure 6.4 Replicated logical concurrency for a Laplace equation solver and accompanying decomposition.

The process structure can also be cast to convey one other possible embodiment of the Laplace equation solver. Maximum logical concurrency can be represented by pre-assigning -- initializing -- each process peer with the required loop indices, and then allowing all $(n\text{-}2)*(m\text{-}2)$ processes to execute. The structure graph for this implementation is simply represented as a large collection of processes, all conducting the same computation, but acting on separate elements of the computational domain, and thus communicating neighboring values between them when needed. All processes in the structure may operate on a small collection of individual elements $U^{new}{}_{i,j}$, and depend on peers for data to complete. In this vein, the computation is logically completed in a simultaneous fashion for the entire domain, on a per-iteration basis.

Figure 6.4 shows this representation as a collection -- an ensemble -- of $N*M$ processes, and each has responsibility for evaluating $U^{new}{}_{i,j}$ for each pair indices independently of the other peers in the structure.

In contrast, Figure 6.3 illustrates a *fine-grain* process structure; each process is nearly decomposed to the atomic operators of input, output, and assignment.[49] Fine-grained process structures are often found in conjunction with SIMD-class multicomputer simulations. In SIMD-class multicomputers, tens of thousands of processors are harnessed to operate on a large dataset by performing relatively simple arithmetic and logical processes in complete synchrony. For a discussion of SIMD-class multicomputers, see Hillis [Hil86].

A *coarse-grain* process structure is illustrated in Figure 6.5. The internals of the replicated processes are hidden and each executes the identical sequential algorithm. Collectively, the processes are joined together with bidirectional communication interfaces (the dual-headed arrows denote this). The replicated logical concurrency in this figure has order eight for processes, and order 16 for communication: eight processes with eight bidirectional communication interfaces. The bidirectional communication interfaces are used by the peer processes to exchange data at some point during their computations. Therefore, each component instance performs many more operations on a larger portion of a partitioned dataset. Expressing more processes, and continually partitioning a dataset among them, implies that each process acquires a smaller responsibility for the overall completion of the computation. A fine-grained structure emerges when the instruction count for each process becomes less than the total number of process instances.

6.2.1 Replicated Logical Concurrency

The large-scale[50] duplication of a structure graph is permitted if a process structure possesses a *replicated* logical concurrency (RLC) attribute. A replicated process structure allows multiple copies of the graph to simultaneously exist and execute within the same physical context (machine or platform). A process structure graph with replicated logical concurrency of order N implies that N copies of the process are running at the same time on a single platform. Each replication acquires independent responsibility, in a computational sense, for a specific region, area, or volume of the computational domain. RLC is vitally important for multicomputer systems. This notion provides the vehicle for realizing maximum speedup from a multicomputer simulation.

Figure 6.5 shows an arbitrarily chosen representation of RLC for the process structure outlined in Figure 6.3. This figure describes the replication of 8 identical Laplace equation solvers as eight identical contexts. The granularity of the Laplace solver process structure shown here is coarser than that of Figure 6.1, in that the individual algorithmic elements are now absent from the picture. RLC hides the

---

[49] Assignment includes statements like $c = a + b$, where arithmetic and/or logical operators are combined and written to a single variable.

[50] Each additional process context consumes memory for active code and data. Efficiency considerations like the granularity of data and message size preclude infinite replication. A maximum of 20,000 to 100,000 process replications are currently possible.

underlying computation structure through encapsulation and insulation. The hierarchical nature of process structure graphs conceals internal details, but highlights the principle units -- processes and communications.

Earlier in our discussion, the focus was placed on the internal computation structure: the organization of individual algorithmic components and operations. They were cast as a collection of processes, which was reduced to one. This yielded a completely sequential iteration of the Laplace equation solution. Encapsulation of the process structure through a collapse of logical concurrency to unit value forces the replacement of a fine-grained representation with a coarse one. Each instance of the algorithm is assigned an independent context, and communication between them occurs at specific -- but unstated -- interfaces. This is the *insulation* of process peers. The process internals of each replication are insulated from its peers by the communication interface.

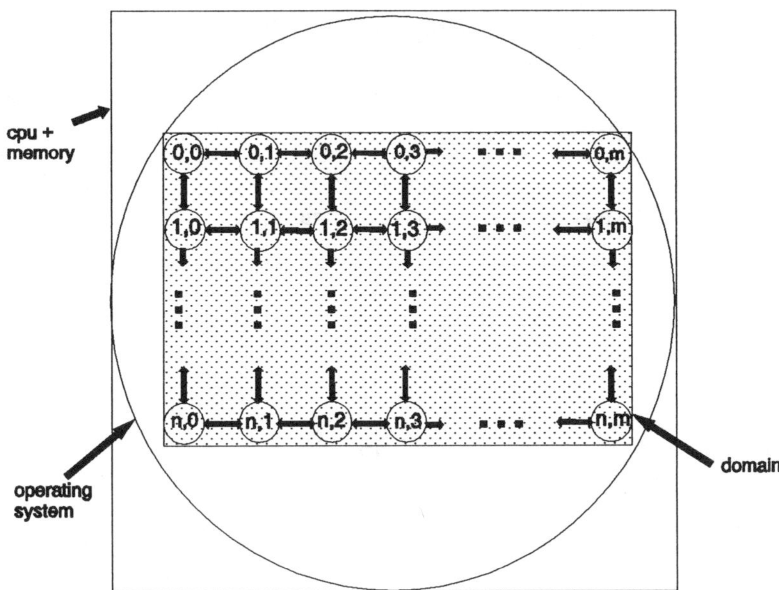

Figure 6.5 An alternative representation of the Laplace process structure graph.

The communication interfaces join processes along one dimension, thus forming a line of processes, and a partitioning of the domain along one axis of the geometry. The arrangement of processes described in Figure 6.5 defines the process structure topology, an important aspect of multicomputer simulation design and implementation.

6.2.2 Process Structure Topology

The process structure *topology* expresses the direction and defines the shape of communication patterns between process peers. The communication pattern shape constrains the volume of message traffic. The shape limits the number of messages -- in a physical sense -- by focusing all communication through a specific set of channels, and is a design feature of the process structure topology. Information is exchanged between processes that comprise the topology, and the process structure explicitly shows this. The process identifiers in Figure 6.5 (the integer labels assigned to each bubble) declare that process 1 exchanges only with process 2, while process 2 exchanges with both 1 and 3, etc. The line of processes described here will ultimately be replicated and joined into this shape during execution.

The process structure topology conveys a logical and geometrical quantity, but necessarily a physical one. The direction of information flow, and the connection of the communication interfaces must be preserved when the process structure is mapped into a physically concurrent computer system. The computer system that most closely matches the process structure topology will execute the simulation most efficiently, but the physical multicomputer configuration may not drastically influence the resulting algorithmic computation. Message-passing operating systems and environments for multicomputers support the mechanisms needed to efficiently map an arbitrary process structure topology onto a congruent physical configuration.

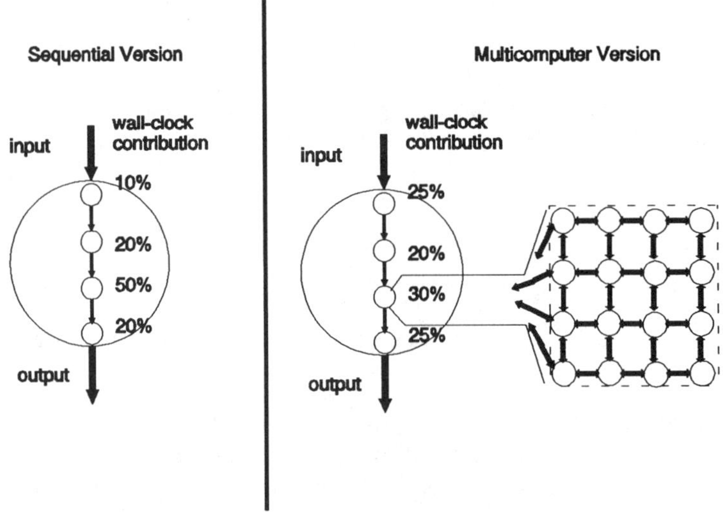

Figure 6.6 Algorithmic structure and qualitative wall clock distribution for sequential process structure and a sequential structure with an embedded multicomputer context.

One selects or designs a process structure topology by considering two independent quantities: the algorithmic structure and the computational domain structure of the problem. The *algorithmic structure* defines the type and organization of each independent function or class used to model or simulate a phenomenon. The simulation component that consumes the most processor resource, as determined by profiling a sequential version of the simulation, is often the most probable candidate for implementation on a multicomputer system. The process structure topology thus becomes altered as Figure 6.6 shows.

The wall clock contribution of each individual component is altered when a concurrent component is added, substituted, or replaces a sequential element in the process structure topology. The dominant resource consumer (identified from a profile of the sequential simulation) has been mollified in a logical sense. Figure 6.6 illustrates how a coprocessor system (in this case a multicomputer) is used to off-load burdensome computations to a dedicated resource. This new process structure topology is appealing at first glance, since the computational burden is eliminated, as indicated by the new wall clock distribution (these are arbitrarily chosen factors for illustrative purposes).

But the benefit depicted by the new process structure topology can be illusory. A huge I/O bottleneck can develop if a substantial amount of data must be shifted into and out of the concurrent element of the process structure for use by the remaining sequential elements.

Under these circumstances, the path between the concurrent and sequential elements is not balanced; the wall clock distribution may worsen. An expedient I/O pathway must be supplied to and from the concurrent context to realize an optimal reduction in the wall clock time. The process structure may include physical mass storage, such as small disks holding 40 to 100 Mbytes each, which can alleviate the overhead associated with access to a single disk farm located at the topology perimeter.

The I/O bottleneck is a central theme to all computer systems; multicomputers are not immune to this artifact, but are subject to an intensification due to their architecture. In many ways, the I/O of a multicomputer resembles that of a light bulb. The light bulb emits a diffuse radiation pattern of approximately equal intensity in an isotropic manner. In contrast, a shared-memory multiprocessor possesses an I/O capacity reminiscent of a laser beam. An intensely concentrated stream is emitted from a single source.

Recasting an existing sequential algorithm into a multicomputer version can be arduous. Fast Fourier transforms (FFT), linear system analysis via factorization, sparse systems analysis via conjugate-gradient, eigensystem analysis, and even matrix multiplication algorithms only resemble their well-known sequential implementations in function, not in appearance and design, once they are transformed to a concurrent context. Progress in this area of parallel computation research has been steady, with many researchers reporting substantial improvements for popular numerical and general-purpose algorithms (see [Fox88a] and [Jam87]).

The *computational domain structure* is the geometrical shape described by the boundaries, internal interactions, and dependencies between the elemental data and objects operated on by the algorithmic structure. The geometry may be quite simple and symmetrical, such as a rectangular region, a linear shape, like a bar, or even the familiar cylindrical and spherical coordinate systems. Geometries that describe simple symmetrical regions are termed *regular*. They present virtually no difficulty for load balancing, and are easy to decompose.

Regular computational domains are useful for two purposes. From an educational standpoint, they provide a convenient framework for training. All physicists-in-training study the solution of classical partial differential equations on regular domains. Regular domains permit simple methods for analytical solution, and provide an avenue of insight into basic physical laws of the universe. A numerical solution to a partial differential equation for a regular domain can be easily checked via analytical means.

Secondly, and most important from a multicomputer simulation perspective, regular domains are easily decomposed and distributed throughout a multicomputer architecture. A regular structure can be divided along a line, or lines, of symmetry that segregates separate regions of computational responsibility. Each region internal to the outer domain boundary that contains the original regular geometry acquires new boundaries when it is decomposed. Information is communicated to neighboring regions along the interfaces between the new boundaries.

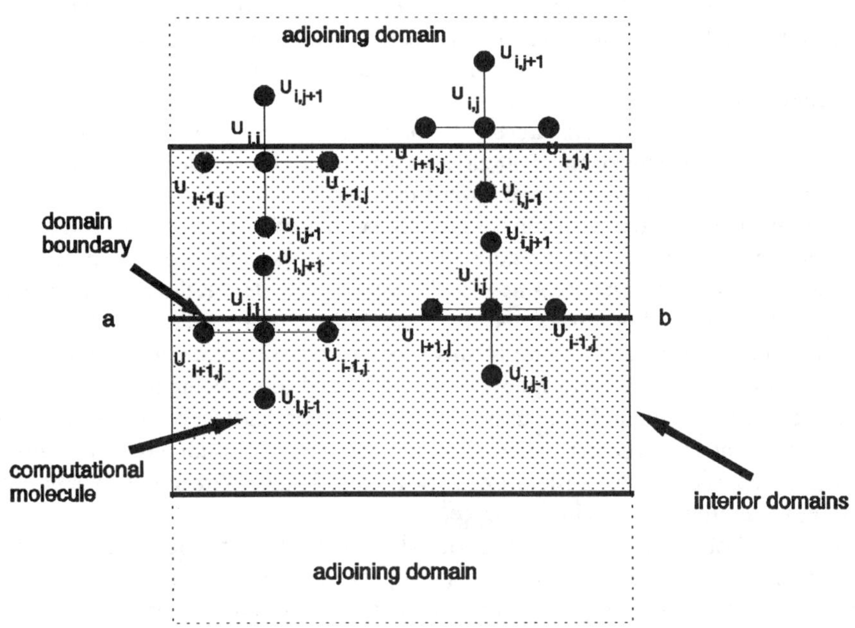

Figure 6.7 Computational molecules at the boundary of internal adjoining domains.

Figure 6.7 shows the computational molecule for the simple finite-difference scheme used to compute the second-order partial derivatives. At the boundaries of the divided domain, the molecules protrude into neighboring domains, and it is at these points specifically that data must be exchanged to complete the derivative computations.

This is illustrated by the Laplace equation solution in Figure 6.5. The RLC clearly shows processes connected to each other along one dimension of the domain. The process structure topology was chosen -- preordained -- to break the original rectangular domain along one dimension. One usually chooses a topology that most efficiently matches the computational domain structure. Rather than a line of processes, one could easily express a two-dimensional mesh or grid of processes. The process structure topology must match the physical structure of the multicomputer system.

From an implementation standpoint, the number of available processors may place constraints on the choice of process structure topology. But the computational domain structure imposes the most visible constraints on the multicomputer implementation. Matching the multicomputer topology to the computational domain structure is the most challenging aspect of multicomputer software implementation (see Chapter 7). In the case of highly symmetric domains, the issue is quite simple, as the computational domain is easily decomposed and distributed. Visualizing a one-dimensional decomposition, like that shown in Figure 6.5 is not difficult. And even two or three-dimensional decompositions are straightforward for symmetric structures.

6.3 Message-passing Basics

Processes exchange data through message-passing. This mechanism is an artifact of computer networks; a stream of data moves from point A to point B. Message-passing subsystems which are embedded within a multicomputer platform distinguish this architecture from distributed computing structures that incorporate local area networks (LAN), wide area networks (WAN), or combine heterogeneous computation platforms into a collectively cooperating edifice. On shared-memory multiprocessor platforms, message-passing among processes can be achieved through several mechanisms, such as pipes, sockets, or shared-memory. But multicomputers are configured with specialized software structures to assist data transfer between processes, and they possess equally efficient hardware support to convey data between physical processes.

Message-passing subsystems implement very efficient mechanisms for moving data streams between processors. The data streams are often controlled by independent direct memory access (DMA) controllers. DMA is an extremely useful method for pushing data into an external device from a memory subsystem. One of the best, if not the finest (and earliest), example illustrating the efficacy of DMA controllers employed for message-passing is the transputer.[51] The transputer employs multiple (four to be exact) independent DMA engines which implement message-passing between neighboring transputers. And the DMA engines can convey the data streams at 20 Mbits/s in a full-duplex mode across four pathways simultaneously. This volume of data transfer amounts to 8 Ethernets-worth of traffic. A modest sum by today's standard, but outstanding when the silicon was first

---

[51] The SGS-Thompson/Inmos T805 transputer @ 20 MHz clock.

introduced in 1984 [Ste88].[52]

6.3.1 Properties and Definitions

MIMD-class multicomputers execute contexts that are intrinsically asynchronous. Processes run at their own pace, and they exchange messages only when needed. For two processes to exchange a message, one of them must send the message, while the other must receive it. While the processes execute asynchronous contexts, they must enter into a synchronous state while exchanging the message data. The two peers remain in a synchronous condition until the message has been completely exchanged. The processes reengage their asynchronous lives only after the message transaction is concluded.

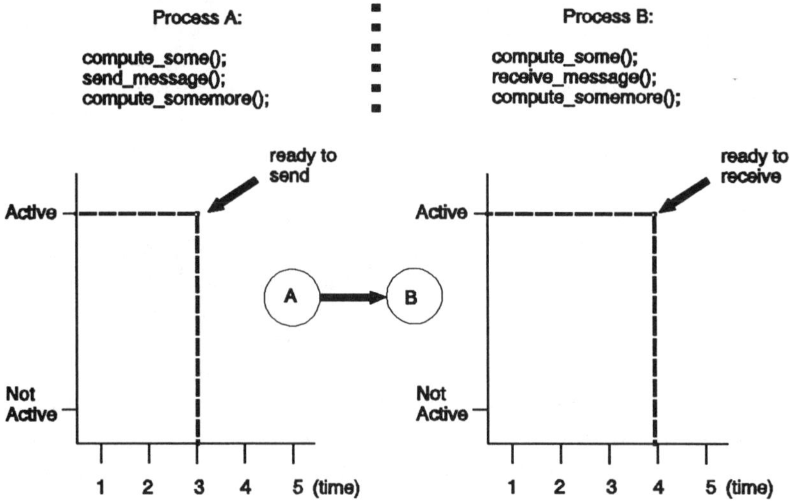

Figure 6.8 Two process peers involved in message-passing.

Figure 6.8 illustrates the timing relationship between two process peers that must exchange a message as part of their functionality. Several items in this figure are noteworthy. The two processes involved here do not reach the message-passing instruction streams at the same time. Process A reaches this point in 3 time-units, while Process B becomes ready in four time-units. Process A must block until B becomes ready to complete the operation. When Process A blocks, no other parts of the instruction stream are executed.

---

[52] The author believes that the transputer is the single greatest British invention since William Shakespeare.

When Process B reaches the message-passing instruction 1 time-unit later, Process A then sends the data to Process B, and they then continue to perform some calculations. The synchronization of the two processes at the message-passing phase of their contexts is vitally important for the preservation of a *deterministic computation*. A computation is deterministic if the processing sequence and outputs are identical each time the same initial conditions or inputs are applied.[53]

Determinism cannot be assured between two process peers, or *N* process peers in general, if synchronization is not possible during message-passing. To see why synchronization is necessary, consider the case where two processes execute simultaneously on two separate processors that share the same address space. Suppose both processes wish to share some variable, call it A_SHARED. Suppose further that one process, Process 1, only requires write-access, while Process 2 requires read-access only. If both processes concurrently execute in parallel without arbitration, what will be the value of A_SHARED at any time? A_SHARED will be indeterminate, since each time Process 1 attempts to write it, Process 2 attempts a read operation at the same time. Thus the value held by A_SHARED will essentially oscillate, and not reach a consistent state. A collision between the processes results from the fact that each process executes asynchronously, so one cannot know when A_SHARED was written.

In message-passing, the same ideal holds true. If Process A sends a message to Process B before it reaches the message-passing instruction to receive, the transmitted information would be lost since Process B has no knowledge of when to expect a packet. Asynchronous message-passing resembles a trial-and-error method of communication among processes. If no synchronizing entity is available to coordinate transmissions and receptions, the processes do not know when to expect data, and it will fall into the bit-bucket, which leads to a non-deterministic situation.

Each process depends on data from peers to complete its role in the overall computation. Obviously, if some or all of the messages are not properly exchanged, or they are entirely absent, a simulation will not progress. Processes communicating via asynchronous message-passing protocols will complete send/receive transactions when they happen to meet at the right time. Probabilistic message-passing encounters provide no certain outcome.

A non-deterministic simulation can also arise if the data required by one process is accidently or mistakenly sent to another. Figure 6.7 clearly indicates the requirement for data to be transferred between nearest neighbors. But suppose the software that specifies the topology of message-passing between nodes 1, 2, and 3, with node 2 responsible for communicating information along its common shared boundaries (1 and 3) mistakenly sends information to the wrong shared boundary; information destined for node 1 from 2 goes to node 3 instead, and information intended for node 3 goes to node 1. Although the process peers are satisfied communication-wise, the information they receive is totally erroneous. In this case, the simulation will run, but the computation will not execute correctly, and will likely produce fallacious results.

---

[53] See [Hoa85] pg. 92 for a formal definition of a deterministic process.

This scenario can occur with great frequency unless the software engineer patiently verifies that the process structure topology is correct, and more importantly, that the software embodiment of the structure is satisfied by the implementation. Automatic formal verification of directional connectivity for the process structure graph is not provided, nor is it ever likely to be except for special cases of highly regular geometries. This operation will rely on human capabilities for the foreseeable future. However, a software design technique is discussed in section 6.5 below which can eliminate much of the guesswork and laborious checking operations.

Enforcing synchronization through the message-passing mechanism imposes determinism on communicating processes. Multicomputer simulations owe their very existence to the secure determinism arising from synchronous message-passing support. Multicomputer OSs that support blocking message-passing primitives introduce the important concept of deadlock to the multicomputer simulation.

6.3.2 Routing

Two process peers that must share information during a simulation may not be physically adjacent or resident on nearest neighbor processors. The process structure topology indicates the logical message-passing interfaces between peers, but does not specify their physical connectivity for the multicomputer platform. Within a multicomputer message-passing network, the adjacent processors hosting processes are *nearest neighbors*. Nearest-neighbor processors are physically connected to each other, and processes resident on them can easily exchange messages between the two physical address spaces. When messages must traverse through an intermediary process, a message *through-routing* scheme ensures that packets reach their proper destination, no matter how many intermediate nodes they pass through on the way.

The earliest single-chip multicomputer nodes did not intrinsically support a through-routing scheme, one that would automatically forward a packet on to the destination. Their architecture supported *point-to-point* message-passing between nearest neighbors. Data is directly transferred between nodes that are hardwired together along a fixed path. Therefore, to realize message-passing between processes resident on non-adjacent processors, a *store-and-forward* implementation of through-routing is necessary. In store-and-forward message-passing, a packet is passed to an intermediate node, a routing table is consulted, and the packet is discharged to the next node on the route. This process requires that the packet be copied into the address space of each intermediate. The earliest versions of message-passing OSs could dynamically allocate buffer space for store-and-forward operations, and then deallocate it after a packet left the processor's address space.

Store-and-forward routing consumes processor cycles and memory resources. But an operating system that supported any form of through-routing on the earliest silicon hosts provided substantial advantages for simulation design and implementation over those that did not.

The latest silicon, such as the Intel iWarp Cell [Kun90], and the proposed T9000 transputer from SGS-Thomson Inmos ([Pou90] and [Pou91]), is either equipped with hardware circuitry support for through-routing which completely alleviates the need for any store-and-forward message-passing, or the chip-set provides a dedicated through-routing system. Both the iWarp and T9000 chip sets support a *packet-switched* through-routing message-passing subsystem. In packet-switching, each packet is prefixed with a code interpreted by the message-passing system operating under microcode control in real-time to determine which circuit the packet should follow to reach a given destination. This technique is almost analogous to the store-and-forward technique, but the packets are not copied into address spaces; they are literally switched between circuits in real-time.

The multicomputer packet switching *circuits* are constructed with the aid of a tool that specifies network connectivity. The grammar is used by the software engineer to identify circuit pathways and construct a connectivity specification. This specification is loaded into the packet-switch network hardware before the simulation begins execution. The packet-switching hardware acts like a telephone exchange network center, and tests to see if a circuit is busy. If so, an alternate path is chosen, and the packet is then routed towards a destination over the least busy or first available circuit. Network hot-spots can be mollified by using a random routing scheme without substantially penalizing efficiency.

From the lucid and practical hypercube performance simulation study of Lamanna and Shaw [Lam91], important and useful models of message-passing performance and behavior are set forth which illuminate interprocessor communication properties.

> "For single packet transmission times between nearest neighbor processors, the following model is assumed for a continuous burst message packet:
>
> $$T_{SL} = t_l + L^*t_c$$
>
> T_{SL} is the single link message delay time to communicate a packet of length L bytes over 1 interprocessor link. T_l is the message latency, which arises from hardware message-passing setup requirements, and t_c is the time to communicate one byte [Lam91]."

In the case where multiple links must be used to forward messages between a sender/receiver pair, the following extension to the single link is used [Lam91]:

$$T_{ML} = t_l + H^*L^*t_c + I^*t_i$$

Where T_{ML} defines the multiple link message packet delay. The quantity H is defined as the number of hops the packet must make within the switching network between the sender and receiver. The quantity I represents the number of intermediates nodes the packet visits for store-and-forward purposes. The quantities H and I are closely related, since the number of intermediate nodes the packet visits between sender/receiver is 1 (one) less than the total length of the communication path.[54]

---

[54] [Lam91] experimentally measure and determine t_l, t_c, and t_i for the Intel iPSC hypercube.

These simple relationships can be adjusted to account for more sophisticated message-passing network structures and methods. Above all, these equations accurately quantify the mathematics of message-passing throughput. They can be used to analyze and predict communication requirements and message loads for a well-defined simulation.

6.3.3 Casting

A process may have a requirement to share information among all peers in the simulation, and not just nearest neighbors. In this case, the process may wish to *broadcast* a packet to all process peers. Packet broadcasting is useful when certain quantities, such as an error value, convergence criterion, iteration count, or other quantity must be globally shared by all processes. The broadcast mechanism implemented for store-and-forward message-passing systems can be expensive if hardware support is unavailable.

The broadcast packet *spread* can be efficiently organized in a binary tree or exponential growth structure. The processor originating the broadcast may contact two nearest neighbors, who in turn contact two more, etc.; a binary exponential growth in the spread is quickly achieved. Johnsson and Ho [Joh89] have exhaustively derived and analyzed routing metrics for broadcast communications in hypercubes.

In addition to broadcasting, the *multicast* and *multireel* functions perform a selective or constrained broadcast. These message-passing operators limit the extent of the broadcast. The multicast routes a single message to a select group of destinations. Multireeling is the converse of multicast: a group of nodes output a message to a single destination. Multicast and multireel operations are often implemented as a sequence of atomic message-passing operations. Both cast operations are analogous to those found in other network structures, such as Ethernet.

6.4 Deadlock[55]

When one process peer attempts a message-passing operation, another process that executes the converse operation must eventually become ready, or a state of *deadlock* is said to exist between the peers. Deadlock prevents the simulation from evolving, since message-passing and all subsequent actions cease until a companion becomes available to complete the transaction. Synchronization serves a double-edged purpose by generating determinism within the simulation, but can also stop a simulation dead in its tracks.

Deadlock frequently arises from faulty software design, and is closely linked to the process structure topology. Process peers are specified with a structure graph that includes their communication interfaces, and software is designed to execute the computation. But if the software is incorrectly organized to account for all interprocess communications when synchronous message-passing primitives are used, then deadlock will surely arise.

---

[55] The scourge of multicomputing!

When a deadlock situation occurs, one may not immediately know where it is located. The frequency of deadlock in a multicomputer system can be diminished if the simulation software can be executed within a single address space, according to the original process structure and logical concurrency specification. While the process structure will not demonstrate the intrinsic speedup when run in this fashion, the single address space software development methodology provides a very expedient device for investigating and verifying if the message-passing will be deadlock-free. Avoiding deadlock in the multicomputer is easy if one is assured that the message-passing operations are matched among process peers.

6.5 Debugging

Debugging a multicomputer simulation is a two step process. Firstly, one must verify that message-passing primitives do not induce deadlock. Secondly, one must verify that the model executes correctly, that the simulation results are well-behaved, rational, and deterministic. The simplest way to verify that message-passing primitives are deadlock-free for a multi-process simulation is to simulate the execution of the physically concurrent process structure.

Physical concurrency refers to the physical configuration of the host multicomputer, and the number of processes the platform will execute when it is loaded with a simulation. Thus, a physically concurrent process structure describes the distribution and topology of the logically concurrent process structure throughout the multicomputer platform. Simulating the physical concurrency implies that the replicated logical concurrency is run in a single address space as multiple, simultaneous processes that communicate through a message-passing interface. A single address space is free of through-routing, packet-switching, and other edifices that contribute to the eventual complication of using a MIMD-class multicomputer.

The Occam model of programming intrinsically provides single address space simulation capability, which is one of its most -- if not the most -- powerful language features ([Wex89] and [Ste88]). In this model, the RLC is organized to reflect the process structure topology, the processes are loaded into memory, and each process executes simultaneously, through timeslicing, as though it were located on a physically distinct processor. Either the operating system, or a microcoded process scheduler timeslices each logical process peer until the computation is completed. When executing in a single address space, the message-passing interface conveys information between processes through *soft-channels*. A soft-channel is a memory-resident one-way, point-to-point communication path, much like a pipe under UNIX. Figure 6.9 illustrates the simulation of a physically concurrent multicomputer simulation via its logically concurrent representation in a single address space using soft-channels.

Soft-channels can be implemented in several ways. If the host computer system happens to run the UNIX operating system, sockets and user data protocol (UDP) datagram services can be used to construct a so-called *stand-alone router* (SAR) ([Ste91] and [Tro90]). The SAR permits multiple processes to communicate via message-passing primitives that are native to a multicomputer silicon host under the host operating system environment. In the transputer domain, a soft-channel is supported by the instruction set, which makes this form of interprocess communication very efficient.

SAR support is used within the confines of a host OS to emulate the physical multicomputer. A SAR library supplies message-passing primitives, and a stand-alone version of the host multicomputer operating system works in conjunction with the SAR library. The stand-alone operating system (SAOS) acts as the hub of a message-passing system in the host computer's address space. Processes are located, in a logical sense, along the circumference of wheel connected to spokes, and communication occurs along the spokes to the hub. The SAOS acts like a centralized controlling process -- a traffic cop -- of the host operating address space transferring packets to peers via the datagrams.

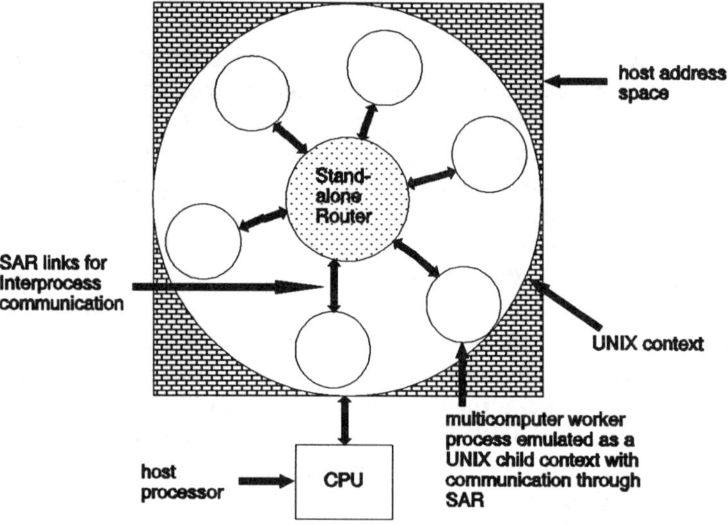

Figure 6.9 The Trollius SAR and multicomputer simulation software emulation architecture.

The essential difference, when using the Trollius SAR to debug message-passing functionality, from an implementation standpoint, is that destination identifiers (node addresses or numbers) for messages are not used to specify routing. Rather, other message attributes, such as event numbers and message-type fields, are needed to route the message from source to destination (see Figure 6.9). This artifact of SAR software support arises from the datagrams and sockets of the UNIX OS used to make the SAR work. This discrimination mechanism imposes an extra minor responsibility on the software engineer. One must be careful to readjust the message-passing subroutine invocations to reflect the difference between the SAR implementation, and the simulation destined for the full-up multicomputer.[56]

The Trollius SAR library and process support software can be used with standard debugging tools, such as dbx. One compiles and links the physical concurrency simulation software with a library that emulates the message-passing functionality of a multicomputer network. The simulation software can be compiled along with standard options for debugging. In this manner, the entire simulation can be single-stepped in the comfort of a single address space by a standard debugger. This technique should be contrasted with the difficulty of debugging a full-up physically concurrent multicomputer, and poking around many separate address spaces to investigate deadlock or some other problem.

Where SAR methodology is not available, a typical approach to debugging a multicomputer simulation is to compile the software with the message-passing library, and bootstrap the simulation software onto the multicomputer. When deadlock or some other problem is discovered, a debugger is invoked to probe around the node address spaces to discover what instruction caused the simulation to halt. If deadlock is the problem, the address spaces of process peers must be examined to isolate the communication problem source. For a data parallel simulation, the remote debugger probe-process can be quite simple provided the computational domain is dependent on nearest neighbor communication patterns. Domains with non-regular communication that involves message through-routing can complicate the situation, and other tools are needed to support the debugging effort.

Execution profilers provide a window on the execution and run-time communication patterns of multicomputer simulation. These tools aid the diagnosis of communication patterns, problems, and load imbalances through a visual user interface. The profiler available with the Express Operating System from the Parasoft Corporation[57] is equipped with a user interface that shows function calls and message-passing traffic. More tools of this type are under development by universities [DMC90] and [DMC91]. The author prefers the SAR methodology, as this technique is intrinsically supported by transputer hardware and the Trollius implementation under UNIX is equally friendly to use. Developing a multicomputer simulation requires practice, and the more familiar one becomes with a set of tools, the easier the effort.[58]

The Trollius[59] [Tro90] SAR emulation of soft-channels is completely

---

[56] A conditional compilation option makes this responsibility quite painless.

[57] Located in Pasadena, CA. Their products are derived from the initial work of the Caltech Concurrent Computation Program started by Geoffrey Fox in the early 1980s.

[58] There is more than one way to skin a cat. This homily applies equally well to multicomputer simulation debugging.

[59] Trollius is commercially available from Transtech Parallel Systems Corporation (Ithaca, NY) as the GENESYS package.

congruent and equivalent to the mechanism supported by the Occam model of programming and Occam compilers. Stand-alone routers are not widely available for other multicomputer silicon hosts, but they offer a terrific advantage for building simulation software. This implies that programming and debugging software on platforms that do not support or embody SAR methodologies are substantially more frustrating and difficult for beginners.[60]

One limitation to the SAR methodology as currently implemented, albeit a minor one, occurs if the physical concurrency is too big to simulate properly. If the simulation software describes a problem that is so utterly complex that it cannot be effectively emulated in a scaled-down fashion with SAR methodology, then one may have no choice other than a full-up multicomputer implementation. However, these cases[61] are quite rare, and usually limited to coarse-grained applications like weather simulation. The SAR methodology should then be used to verify message-passing functionality when the problem is too coarse. Stub computation routines are inserted to substitute for the actual number-crunching algorithms.

A second limitation with the SAR implementation of Trollius 2.0 is that it does not permit either broadcast or multicast message-passing. A work-around solution for this limitation might be a sequential series of message-passing calls, one to each peer process.

The goal to achieve with SAR methodology is the eventual installation of the logically concurrent process structure onto the physically concurrent multicomputer system. Once the simulation is loaded, one can expect processing to occur at the realizable limits of the platform. SAR methodology makes this process painless and relatively swift, especially if compared with the shotgun alternative.

6.6 Physical Concurrency and Multicomputer Topology

Once the physical concurrency emulation via the SAR methodology is shown to be deadlock-free, the software is ready to map into the multicomputer for execution of the logical process structure. SAR methodology eases this transformation to physical concurrency. One must relink the simulation model with a different library of message-passing routines that preserve the name and calling interface convention used in the SAR version. Secondly, the node address message-passing convention must be substituted for the message event and type attributes required by the SAR implementation. At this point, the simulation is almost ready for launching into the multicomputer platform.

The final preparations are quite simple. The specification of the physical multicomputer topology is needed to inform the routing software of the physical node connectivity to preserve the message-passing structure described and required by the process structure topology. A node and interconnect language (NaIL) is supplied with

---

[60] This is the author's opinion.

[61] Where SAR methodology is unavailable, the simulation is shotgunned onto the multicomputer, and an equally appropriate debugging methodology is attempted.

the Trollius Operating System [Tro90] for this purpose. A file is organized that specifies how the nodes are connected. When this description is completed and the specification has been assembled by a utility program, a ***hard-channel*** is established through which the messages between processes will pass. A hard-channel is the hardware implementation of a soft-channel. The soft-channels are transformed to hard-channels through the relinking process.

The hard-channel connectivity information is established through a NaIL specification. This specification is fed into a utility which loads the multicomputer OS with the routing information and tables used to direct messages during execution.

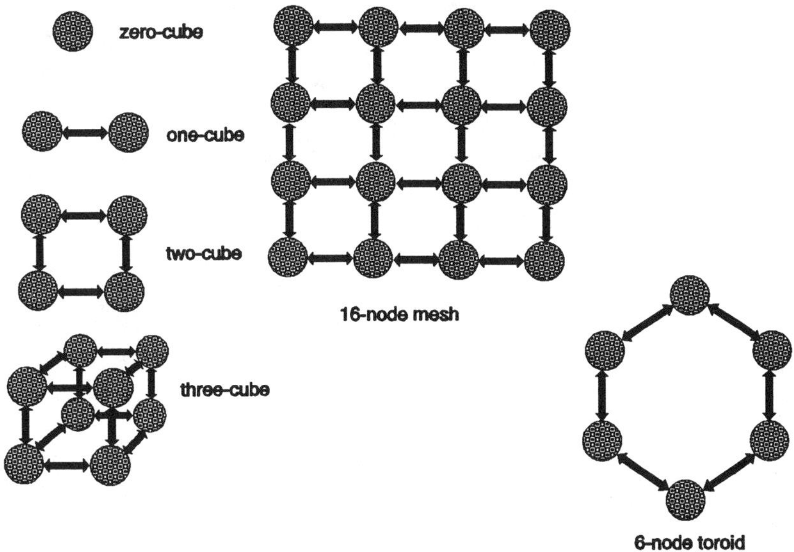

Figure 6.10 Common multicomputer topologies.

One chooses the physical multicomputer topology to optimize a particular simulation structure. For example, *N-body problems* require the evaluation of forces and interactions among all particles or objects within the domain against all others. Molecular or particle dynamics equations of motion depend on the computation of N forces against (N-1) particles, and this is most easily accomplished with a toroidal multicomputer topology. Figure 6.10 illustrates a few common physical topologies.

A hypercube topology possesses the most efficient point-to-point communication configuration available for message-passing across the diameter of a multicomputer message-passing network. Many commercially available multicomputer platforms are equipped with hypercube wiring topologies; they provide the shortest physical interconnection distance between corners of the cube. The hypercube platforms allow the fastest message-routing since the minimum travel distance for signals is afforded by the cube configuration.

The machines wired into a hypercube interconnection structure often permit lower-dimensional topologies to be extracted and configured to match a particular computation domain structure. This process is aided with a multicomputer operating system, such as the Reactive Kernel/Cosmic Environment of the California Institute of Technology Mark II & III hypercubes [Seiz88].

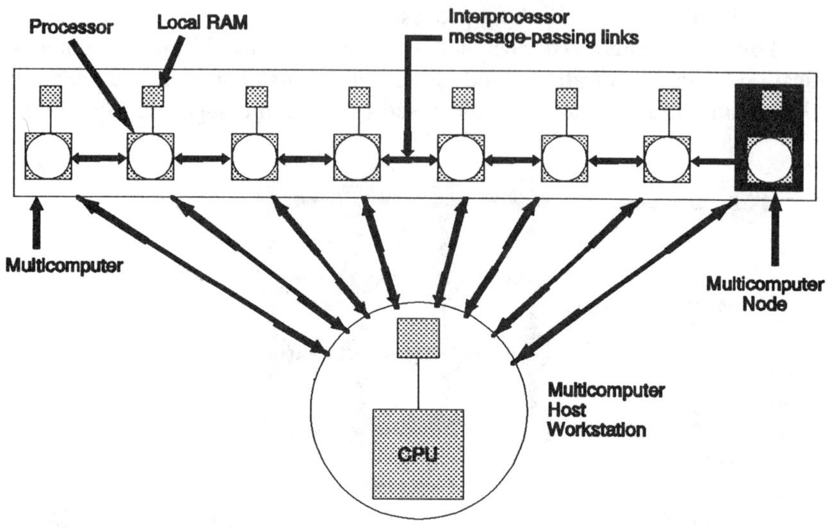

Figure 6.11 Physical multicomputer topology congruent with logical concurrency representation. (Byte Magazine, June 1991 © McGraw-Hill, Inc., New York. All rights reserved.)

In the case of the logical concurrency expressed by the process structure topology of Figure 6.5, the physical concurrency of the multicomputer topology assumes the configuration in Figure 6.11.

6.7 A Numerical Design Example

In this section, a simple two-dimensional regular domain is used as a canvas for illustrating the techniques outlined earlier in this chapter. The problem to solve is outlined below [Bur81]. Figure 6.12 illustrates the regular, rectangular geometrical domain (the computational domain), boundary conditions and source function for the example. The problem to solve corresponds to the determination of steady-state heat or flux evolution subject to the conditions shown in Figure 6.12. A Poisson equation solution is described which employs a basic iterative solution via finite differences with successive over-relaxation (SOR) to accelerate convergence of the Gauss-Seidel process. The foundations for this exercise are illustrated throughout the process structure graphs and logical concurrency descriptions given earlier in this chapter.

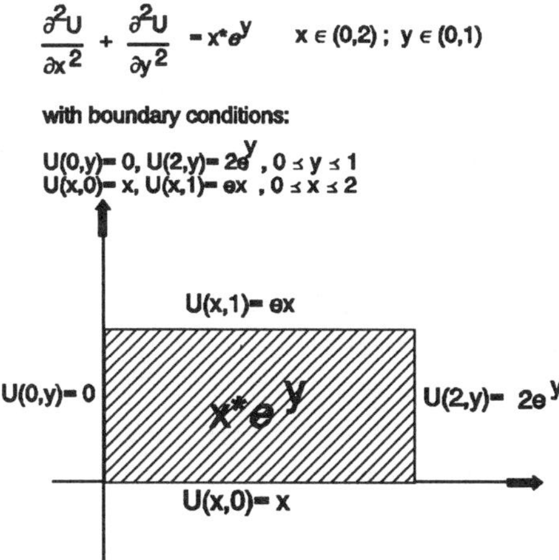

Figure 6.12 Sample geometry and boundary conditions for Poisson equation system.

The solution approach taken here is based on finite differences. Rigorous details can be found in an earlier publication [Ste90].

To begin, the central difference approximation for the partial derivatives are obtained by evaluating the Taylor series at the points $x = x_i$ and $y = y_j$. This yields Equation 6.2:

$$\frac{\partial^2 U}{\partial x^2} + \frac{\partial^2 U}{\partial y^2} = \rho(x,y), \text{ and } (x, y) \in (a, b) \times (c, d)$$

and $U(x, y) = g(x, y)$ *when* $x = a$ *or* $x = b$ *or* $y = c$ *or* $y = d$
For reasonable functions g, this problem has a unique solution. An approximation to the solution can be obtained as follows: for integers n and m, define stepsizes $h = (b - a)/n$, $k = (d - c)/m$, *with* $x_i = a + ih$, $i = 0, \ldots, n$, *and* $y_j = c + jk$, $j = 0, \ldots, m$. 6.2

Combining and rearranging the terms from the partial derivative approximations produce Equation 6.3:

$$\frac{\partial^2 U}{\partial x^2} = \frac{U_{i-1,j} - 2*U_{i,j} + U_{i+1,j}}{h^2} + O(h^2)$$
$$\frac{\partial^2 U}{\partial y^2} = \frac{U_{i,j-1} + -2*U_{i,j} + U_{i,j+1}}{k^2} + O(k^2) \quad 6.3$$

This is the Gauss-Seidel iteration formula. The problem domain we have described gives rise the computational molecules found in Figure 6.6. Solvers like Gauss-Seidel require many iterations to converge. The entire computation, as approximated by finite differences, amounts to an averaging process over the entire domain.[62] Convergence can be improved by using applying successive over-relaxation (SOR), which performs a weighted averaging of previous iterations over the field variable $U_{i,j}$. Equation 6.3 can be rewritten to yield Equation 6.4, where the SOR parameters are combined into the finite difference equations (Equation 6.5).

$$2[\frac{h^2}{k^2} + 1]*U_{i,j} - [U_{i+1,j} + U_{i-1,j}] - \frac{h^2}{k^2}[U_{i,j+1} + U_{i,j-1}] = -h^2*\rho_{i,j}$$

for every $i = 1, 2, \ldots, n-1$ *and* $j = 1, 2, \ldots, m-1$ *and*

$U_{0,j} = \rho(x_0, y_j),\ j = 0, 1, \ldots, m$ 6.4

$U_{n,j} = \rho(x_n, y_j),\ j = 0, 1, \ldots, m$

$U_{i,0} = \rho(x_i, y_0),\ i = 0, 1, \ldots, n$

$U_{i,m} = \rho(x_i, y_m),\ i = 0, 1, \ldots, n$

$$U^*_{i,j} = \frac{[-h^2*\rho_{i,j} + (U_{i+1,j} + U_{i-1,j}) + \mu*(U_{i,j+1} + U_{i,j-1})]}{\lambda}$$

where $\mu = \frac{h^2}{k^2}$, *and* $\lambda = 2*(\mu+1)$

The SOR formula is a weighted average computation.

The SOR average is expressed as: 6.5

$U^{new}_{i,j} = \omega*U^*_{i,j} + (1 - \omega)*U^{old}_{i,j}$

where ω, *the relaxation parameter, is computed as:*

$\omega = \frac{2}{\sqrt{(1 - \rho(U)^2)}}$, *and* $\rho(U)$ *is the spectral radius of the matrix U.*

The SOR parameters are defined by Equation 6.6.

$\rho(U) = \frac{1}{2[\cos(\frac{\pi}{n}) + \cos(\frac{\pi}{m})]}$, *where n, m are the grid quantization factors. The relaxation parameter* $\omega = \frac{4}{2 + \sqrt{4 - (\cos(\frac{\pi}{n}) + \cos(\frac{\pi}{m}))^2}}$.

The following equation describes the standard sequential implementation which one might implement on a personal computer:
To implement this algorithm on an N processor multicomputer, a data decomposition must be accomplished. A portion of the rectangular domain must be placed on each

[62] Conjugate-gradient solutions usually converge much faster.

for $i = 1$ *to* $n-1$, *step* 1
  for $j = 1$ *to* $m-1$ **6.7**
    $u_{i,j} = \omega[u_{i-1,j} + u_{i+1,j} + u_{i,j+1} - h^2\rho_{i,j}] - (\omega-1)u_{i,j};$

processor, and a topology must be chosen to communicate information across processor boundaries.

The simplest decomposition to implement involves the domain dissection along one dimension of the rectangular geometry. An alternative decomposition, and the most natural one to select, is the one which involves decomposition along both dimensions of the geometry. The choice of one dimension here is simply limited by the number of available processors -- 8 to be exact -- which could be molded into a 2 by 4 mesh, but it is simple enough to describe a linear topology via the Trollius NaIL. By choosing this decomposition, we need to loop-split one of the for-loop control structures such that it is parameterized.

Thus, the loop:

for $i = 1$ *to* $n-1$, *step* 1

for arbitrary row-wise decomposition, becomes:

for $i =$ *istart to* $i =$ *istop, step* 1

In consideration of the one-dimensional decomposition, the row or column-wise loop-split generates the smallest amount of message-passing overhead when compared with the two-dimensional decomposition. Also, when the decomposition is formally implemented, the domain parameters, the new geometrical limits on the computational domains formed by the decomposition, become the start/stop indices for the local computation of boundary conditions for each sub-domain created as a result of the decomposition process. The rectangular geometry is represented by each independent element $U_{i,j}$, and this field describes is a full, positive-definite matrix.

Now that the basic decomposition and loop-splitting has been accomplished, the next step towards the implementating the multicomputer version is to simulate the decomposed, loop-split algorithmic version in a single address-space to show that the Gauss-Siedel algorithm remains consistent. The author has a preference for constructing a parameterization of the loop-split operation which mirrors the one ultimately destined for the multicomputer system. By performing this manual loop splitting process, the internal communication boundaries of the multicomputer version become easily visible. The single address space decomposition shown below divides the computational domain into four regions. Two of them, the two internal boundaries created by the loop-split, will require bidirectional communication when placed on a multicomputer. Since this example illustrates a single address space version, no message-passing is needed; addresses are retrieved sequentially. This is performed as follows:

This algorithm was tested in a single address space just to verify that decomposition of the matrix U, and subsequent iteration on it by the loop-split version,

```
1) Organize the data decomposition;
2) Determine where to partition the iterative solution.
for i= 1 to i=n-1 , step1
for j= 1 to j=m-1 , step 1
becomes:
for i= 1 to i=(n/4 - 1), step 1
  for j= 1 to j=m, step 1
    do_gs_iterations(i,j);
for i= nrows/4 to i=(2*n/4) - 1, step 1                    6.9
  for j= 1to j=m, step 1
    do_gs_iterations(i,j);
for i=2*nrows/4 to i=(3*nrows/4) - 1, step 1
  for j= 1 to j= m, step 1
    do_gs_iterations(i,j);
for i=3*nrows/4 to i= nrows - 1, step 1
  for j= 1 to j= m, step 1
    do_gs_iterations(i,j);
```

does not destroy the numerical integrity of the algorithm. Some engineers might regard this step as highly conservative and unnecessary, but the author believes that jumping too many steps during the multicomputer software engineering process, and racing toward the final goal too quickly, invites problems that can be avoided with a little caution and extra preparation. Remember the three d's?!

The next step in the software engineering process involves the installation of message-passing primitives to permit the interprocess communication between peers. Inserting the primitives at this stage allows the process structure topology that was introduced as a result of the data decomposition to emulate the physical concurrency ultimately sought. The SAR methodology is employed for the purpose of simulating the physical concurrency (see the following equation).

For domains interior to the top (the edge along the line $x = 0, y = 0$ to $y = 1$) and bottom (the edge along the line $x = 2, y = 0$ to $y = 1$) boundaries of the geometry, a pseudocode fragment for a region that communicates with two nearest-neighbor sub-domains is shown with the message-passing primitives and the iterative portion of the Gauss-Seidel algorithm.

The quantities *vec1* and *vec2* are used to capture the in-coming messages received from the boundary at the top of the domain held by the processor immediately above the local node (nodeidup) with the invocation *getminefromabove(eventdown)*. Messages received from the boundary of the processor immediately below the local node (nodeiddown) are obtained via the invocation of *getminefromabove(eventup)*. The data to transmit to processor above and below the local node are embedded in the invocations: *sendminetoabove(nodeidup, eventdown)* and *sendminetobelow(nodeiddown, eventup)*. The send operations transmit the vectors *U*_{istart,j} and *U*_{istop,j} for all j to the processors that straddle the local node.

In the case of a node which acquires packets from two nearest neighbor processors, like that explained above, the event signatures *eventup* and *eventdown* are used to discriminate between the incoming packets. Without these filters, which are

```
getminefromabove (eventup) ;
getminefromabove (eventdown) ;
sendminetoabove (nodeidup, eventdown) ;
sendminetobelow (nodeiddown, eventup) ;
for (i = istart ; i ≤ istop ; i++ )
{
  int k= i - istart ;
  for (j = 1 ; j ≤ m; j++ )
  if ( i ≡ istart )
  {
    z= (h²*ρ(x_i,y_j) + u_{k+1,j} + vec1[j] +
       μ*(u_{k,j+1} + u_{k,j-1}))*λ ;                              6.10
    u_{k,j}= ω*z + (ω - 1)*u_{k,j} ;
  }
  else if ( i ≡ istop )
  {
    z= - (h²*ρ_{x_i,y_j} + u_{k-1,j} + vec2[j] +
       μ*(u_{k,j+1} + u_{k,j-1}))*λ ;
    u_{k,j}= ω*z + (ω - 1)*u_{k,j};
  }
  else /* do everything else in between istart and istop */
}
```

used by the message-passing primitives, a definite possibility of incorporating the wrong sub-domain boundary values in the conditional code for i == istart and i == istop exists. Without the event field, a non-deterministic computation will develop.

When two processes that are resident on separate processors send messages to the same node -- the local node, the local node receiving process uses the event field of the packet header to discriminate the packets. If non-unique events are used, the receiving process may employ the incorrect data for a particular portion of the computation. The following situation illustrates this possibility:

```
send(node3, event1); /* from node 2 to node 3 */
send(node3, event1); /* from node 4 to node 3 */
```

The organization of code on processors 2 and 4 will most likely result in an ill-convergent (non-deterministic) computation. In the rare case where the temporal packet ordering is fortuitously correct, the computation will converge, but this is only probabilistic and not reproducible from run to run. The correct usage of message-passing primitives should resemble:

```
send(node3,event1); /* node 2 to node 3 with event 1 */
send(node3,event2); /* node 4 to node 3 with event2 */
```

And on the receiving end, node 2 will contain logic of this form:

```
receive(event1); /* from node 2 to node 3 */
receive(event2); /* from node 4 to node 3 */
```

and this is what our algorithm indicates.

To complete this example, we must embed the message-passing functions into the algorithm. Placing the message-passing invocations in the source code links the independent processes into a fixed topology when the separate processes are loaded into the multicomputer system. This simulation can be executed on N multicomputer nodes which are organized into a linear topology. To obtain maximum flexibility, the algorithm must be parameterized for decomposition among the N nodes. The basic outline of the algorithm follows. The variable *whichcpu* refers to the local node identifier, which are assigned increasing numeric integer values from 1 to N, that sends to a nearest neighbor (nodeidup = whichcpu - 1, or nodeiddown = whichcpu + 1).

The communication primitives used in this code fragment, sendmineto{above,below} and getminefrom{above,below} perform one-way message-passing between the local node (whichcpu), and either the node immediate above or below. Conversely, the getminefrom{above,below} receive packets from nodes either above or below the local node.

$$
\begin{array}{l}
\textit{for}\ (\ i = \textit{istart};\ i \le \textit{istop};\ i{+}{+})\\
\{\\
\quad \textit{if}\ (\ i \equiv \textit{istop}\ )\\
\quad\quad \textit{for}\ (\ j = 1;\ j \le \textit{ncolumns} - 1;\ j{+}{+}\ )\\
\quad\quad \{\\
\quad\quad \textit{double}\ z = (-dx^2 * \rho_{i,j} + \textit{msg\_vector}_j +\\
\quad\quad \textit{solution}_{i-1,j} + \lambda * (\textit{solution}_{i,j+1} + \textit{solution}_{i,j-1})) * \mu\ ;\\
\quad\quad \textit{solution}_{i,j} = \omega * z + (\omega - 1) * \textit{solution}_{i,j}\ ;\\
\quad\quad \}\\
\quad \textit{else}\\
\quad\quad \textit{for}(j = 1;\ j \le \textit{ncolumns} - 1;\ j{+}{+}\ )\\
\quad\quad \{\\
\quad\quad \textit{double}\ z = (-dx^2 * \rho_{i,j} + \textit{solution}_{i+1,j}\\
\quad\quad \textit{solution}_{i-1,j} + \lambda * (\textit{solution}_{i,j+1} + \textit{solution}_{i,j-1})) * \mu\ ;\\
\quad\quad \textit{solution}_{i,j} = \omega * z + (\omega - 1) * \textit{solution}_{i,j}\ ;\\
\quad\quad \}\\
\}
\end{array}
\tag{6.11}
$$

The values of *istart* and *istop* are computed inside the local initalization routine, and are derived from the values of num_nodes, the number of rows, and the number of columns used to quantize the grid to compute the finite difference approximation of Poisson's equation. The routines **compute_top**, **compute_mid**, and **compute_bot** accept parameters as follows:

compute_top(solution, msg_vector, istart, istop)
double **solution, *msg_vector;

int istart, istop;

```
for ( i= istart; i ≤ istop; i++) /* compute-mid */
{
  int k= i - istart ;
  if ( i ≡ istart )
    for ( j= 1; j ≤ ncolumns - 1; j++ )
    {
    double z= (-dx^2*ρ_{i,j} + solution_{k+1,j}+
    msg_vector1_j + λ*(solution_{k,j+1} + solution_{k,j-1}))*μ ;
    solution_{k,j}= ω*z + (ω-1)*solution_{k,j} ;
    }
  else if ( i ≡ istop )
    for(j = 1; j ≤ ncolumns - 1; j++ )
    {
    double z= (-dx^2*ρ_{i,j} + msg_vector2_j +
    solution_{k-1,j} + λ*(solution_{k,j+1} + solution_{k,j-1}))*μ ;
    solution_{k,j}= ω*z + (ω - 1)*solution_{k,j} ;
    }
  else
    for ( j= 1; j ≤ ncolumns - 1; j++)
    {
    double z= (-dx^2*ρ_{i,j} + solution_{k+1,j} +
    solution_{k-1,j} + λ*(solution_{k,j+1} + solutionk,j-1))*μ ;
    solution_{k,j}= ω*z + (ω - 1)*solution_{k,j} ;
    }
}
```

Solution is the local portion of the field *U* organized as a 2x2 matrix (hence the double indirection). *Msg_vector* is used to pass in boundary information to complete the Gauss-Seidel iteration. The data pointed at by msg_vector always arrives from the processor immediately below the head of the line. A pseudocode fragment illustrating the **compute_top** functionality follows. The *istart* and *istop* serve as our data decomposition parameters for the iteration of the local (resident in the address space held by processor whichcpu) field *U* (*solution*).

The **compute_bot** routine is identical in every way to the compute_top routine, except that msg_vector is substituted for $\text{solution}_{i-1,j}$. The **compute_mid** routine requires two *msg_vector* inputs, one corresponding to the data received from the processor above the local node, and one corresponding to the data received from the processor below the local node.

compute_mid(solution, msg_vector1, msg_vector2, istart, istop)
double **solution, *msg_vector1, *msg_vector2;
int istart, istop;

Note that msg_vector1 and msg_vector2 are used within the algorithm at locations that correspond exactly to their geometric dependence of the computational domain structure following decomposition. For each middle section of the decomposed geometry, the messages arriving from processors below the local node are used when i= *istop*, the end of the parameterized loop. The messages arriving from above the local node are used only when i= *istart*. Points to be computed between the boundaries of each sub-domain are only obliquely related to the points located on the boundaries. The *istart* and *istop* parameters have the same definitions as compute_top and compute_bot.

Each node in the multicomputer executes an identical code fragment. That we have chosen to build the topology decode logic into the multicomputer simulation supplies the freedom to run the software on *N* nodes, so long as they afford nearest neighbor communications that emulate a linear topology. The code fragment for the physical implementation follows.

```
/* obtain num_nodes, max. iterations, & other parameters */
/* set istart, istop to appropriate values for position within the line */
/* the line of processors; whichcpu is established at bootstrap for each */
/* processor */
initialize();

while (current_iteration < max_iterations)
if ( num_nodes > 2 )
{
/* if more than 2 nodes in the topology */
if ( whichcpu > 1 && whichcpu < num_nodes )
{
/* if I'm somewhere in the middle of the topology */
  if ( (whichcpu % 2) = = 0 )
  /* If I'm an even-numbered cpu in the topology */
  {
        sendminetoabove(nodeidup,whichcpu);
        getminefromabove(nodeidup);
        getminefrombelow(nodeiddown);
        sendminetobelow(nodeiddown,whichcpu);
  }
  else /* I'm an odd-numbered cpu in the topology */
  {
        sendminetoabove(nodeidup,whichcpu);
        getminefromabove(nodeidup);
        getminefrombelow(nodeiddown);
        sendminetoabove(nodeiddown,whichcpu);
  }
        /* do Gauss-Siedel iterations */
        compute_mid(U, vec1, vect2, istart, istop);
else
/* either at the head or end of the line */
{
```

```
    if ( whichcpu == 1 )
    { /* I'm the head of the line */

            getminefrombelow(nodeiddown);
            sendminetobelow(nodeiddown,whichcpu);
            compute_top(U,vec2,istart,istop);
    }
    else if ( which_cpu == num_nodes )
    { /* I'm at the tail */
            sendminetoabove(nodeidup,whichcpu);
            getminefromabove(nodeidup);
            compute_bot(U,vec1,istart,istop);
    }
   }
  }
  else if ( num_nodes == 2 )
  { /* only 2 processors in the topology */
    if ( whichcpu == 1 )
    {  /* I'm the head of the 2 processors */
            getminefrombelow(nodeiddown);
            sendminetobelow(nodeiddown,whichcpu);
            compute_top(U,vec2,istart,istop);
    }
    else if ( whichcpu == num_nodes )
    {/* I'm the bottom of the 2 processors */
            getminefromabove(nodeidup);
            sendminetobelow(nodeidup,whichcpu);
            compute_bot(U,vec1,istart,istop);
    }
   }
  current_iteration++;
  } /* end while iterations */
```

The approach adopted here is termed same-code/multiple-data (SCMD), as defined by Quinn and Hatcher ([Qui90a], [Qui90b], and [Qui91]). The identical processing sequence is replicated and installed on each processor, but operates on a separate portion of the computational domain as assigned by the control logic embedded in the program body. The program body illustrates the physical concurrency of the multicomputer implementation. The line of processors and their computational domain assignments are clearly demarked by the conditional structure and the message-passing structure embedded within each nesting level. The initialization subroutine invoked by each processor identifies its range of spatial computational responsibility based on the geometric boundary constraints.

6.8 Alternatives to Explicit Message-passing

The Linda[63] system ([Ahu86], [Lel90]), developed by David Gelernter, affords a unique, flexible, and portable parallel programming system which alleviates much of the multicomputer software engineering effort previously outlined. Linda employs a novel structure, the *tuple-space*, to organize and unify memory references in a distributed system, like a multicomputer. The tuple space resembles a blackboard against which data structures and entities can be checked-in, checked-out or copies made for examination and manipulation by a process. The tuple-space operators *in*, *out*, *eval*, and *read* drastically simplify parallel program design [Car88].

Tuple-space is structurally organized over the extent of the distributed memory available within the multicomputer. A tuple-space is named, and references by processes to tuple-space objects and members are possible which may not physically reside within the processor's address space; the tuple-space object is located, and transparently passed to the referencing process. No explicit message-passing operator is needed. Linda arranges for the distribution of tuple-space objects through a hash function. The *in* operator installs data into a tuple-space, where other processes residing in different address spaces can reference the information through the *out* or *read* operators. The *eval* operator evaluates tuple-space components which have a functional structure. Suppose the following sequence of tuple-space operators occured *in*(sqrt(3.14)), **root**= eval(sqrt(3.14)). The atom sqrt(3.14) is placed in tuple-space, and the *eval* operator is invoked and actually computes the value of the atom returned in the variable **root**. The simplicity of the Linda operators, and the tuple-space structure form a powerfully useful mechanism which conceals the platform's architectural dependencies from the user. The user can remain relatively niave about whether the program is running on a shared or distributed memory platform, and construct parallel programs with Linda primitives which are extraordinarily portable.

The architectural features of multicomputer platforms complicate the development of the software. Mechanisms for alleviating these dependencies make the software engineering tasks much simpler, although hardware vendors lose their customers faster. This is the role of the multicomputer operating system. In some ways, the Linda tuple-space mechanism acts as a surrogate parallel random access machine (PRAM)[64]. The logical data references requested by processes are decoded and mapped into physical machine addresses which may reside in disparate locations in the multicomputer. The requesting process receives the data following this decode operation without knowing if message-passing or a 32-bit parallel bus was used to transmit the data.

---

63 Available from Scientific Computing Associates, Inc. New Haven, Connecticut.

64 PRAM technology is discussed in Chapter 9.

6.9 Program Efficiency

A multicomputer's capacity to achieve performance gains -- speedup -- when executing a simulation is balanced with the overall efficiency of processor usage, message-passing requirements, and the synchronization imposed by the message-passing. Flatt and Kennedy [Fla89] have derived a set of metrics which analytically express optimum program efficiency and speedup in relationship to multicomputer synchronization and message-passing penalties.

They show for the early hypercubes developed at Caltech, that for maximum efficiency, "the portion of the computation that can be performed in parallel must be over 99% of the total computation and the overhead function for 64 processors must be small than 3 tenths of a percent of the total computation [Fla89]." Their results prove that a high degree of parallelism and small over-head (e.g. message-passing or synchronization imposed by it) is necessary for top efficiency. When message-passing is expensive, try to use less, and keep the processors busy. This can be achieved by using a coarser data decomposition; placing more data on each processor forces the cpu to perform less message-passing.

Each application is different. Small grained applications, like simple image processing or the calculation in section 6.7 perform few computations at the individual data point level. The grainularity is fine, and the motivation for decomposing the computational domain into fairly large regions is a reasonably strong. Further decomposition of the domain where only a few elements reside on a processor is unwise, since message-passing on early silicon is quite expensive. Alternatively, a weather calculation or model consumes several thousands of calculations per data point. The ratio of computation to communication is huge (of order 1000:1). High-efficiency and an increasing level of speedup arises in this case.

Concluding Remarks

The following steps summarize the design methodology presented here:

(1) Construct a process structure that reflects the computational granularity of the simulation while keeping in mind the requirements for processor resource utilization for message-passing and computation. Choose a granularity that will maximize either the computational bandwidth or the message-passing throughput or both.

(2) Detect the areas of the process structure graph that can be reduced or collapsed to concurrency measure nearest to unity. This may be accomplished by eliminating communications and unnecessary process peers. Use sequential algorithmic components where prudent. Although a processor may be equipped with special purpose instructions and structures to permit many simultaneous processes, use this capability wisely.

(3) Simulate the physical concurrency via a stand-alone router or soft-channel message-passing mechanism. It is far easier to debug a simulation in a single address space than poking around multiple address spaces.

(4) Finally, map the process structure onto the multicomputer.

It should be clear from the discussion presented in this chapter that MIMD-class multicomputer software engineering requires substantially different thought processes to conduct. The only means of acquiring an effective working knowledge of these though processes is to get out and work on a few multicomputer simulations. The transputer provides the most cost-effective means for conducting this exercise and gaining the necessary experience. For about US $500, one can learn concurrency, and master the necessary techniques to build ever more complicated scalable simulations.

One drawback of the SAR methodology the author promulgates lies with the ordering of message-passing primitives in the simulation code. The SAR does not care about the order, since the datagram and socket software it relies on is non-blocking. One send, and one receive is necessary to prevent deadlock in SAR environment. The physical multicomputer cares a great deal about the primitive ordering, especially when singleton communication routines are used.

The CUBIX message-passing library provides bidirectional message-passing routines that eliminate the need for singletons. CUBIX was developed by Caltech, and is supported by both the Trollius OS and Express environment.

Linda should be strongly considered for distributed multicomputer applications, where excess computer cycles are available during off-peak hours [Mar92a]. Linda's tuple-space structure is an attractive alternative to explicit message-passing. The tuple-space operators *in*, *read*, *eval*, and *out* manage data lookup and message-passing operations between processes, thus relieving the user from programming these explicit dependencies. The tuple-space functionality supplies a substantial advantage [Lel90a] for parallel program design.

Quinn and Hatcher ([Qui90a] and [Qui90b]) have developed and experimented with a data parallel compiler which generates a decomposition for use within a MIMD-class multicomputer environment. Their results are encouraging to the extent that the SCMD approach may soon become nearly automatic for many problem domains.

Suggested Reading

The author strongly recommends that the reader acquire experience with the Occam language as a training tool to become educated and acquainted with the principles of logical and physical concurrency, and message-passing. The preeminent textbook on Occam programming is that by Wexler [Wex89]. C.A.R. Hoare [Hoa91] has written an entertaining and enlightening history of the Occam language. A rigorous, though incomplete, survey and evaluation of multicomputer operating systems, tools, and utilities can found in [Cha90]. Fox *et al*. [Fox88a] has written the preeminent text on multicomputer solutions for regular problems. Some of the earliest work on routing in hypercubes can be found in Fox and Furmanski [Fur86]. Bokhari [Bok90] has performed an in-depth experimental evaluation of the Intel iPSC-860 communication overhead.

7

Load Balancing

This chapter gives an overview of load balancing mechanisms and techniques for MIMD-class multicomputers. A multicomputer simulation achieves greatest utility and efficiency when all nodes simultaneously process datasets of equal size. A proper load balance condition is not always easy to reach, especially where irregular domains are concerned; their computational domain structure often precludes simple load balancing via inspection and data decomposition. A properly balanced multicomputer simulation minimizes the dead-time between synchronizing process peers. This condition arises from a computationally unbalanced multicomputer where processors spend more time in a blocked state waiting for process peers to reach the communication stage of the computation.

Several methods for MIMD-class multicomputer load balancing have been published in the open literature, and some of these are described here. The inspection method of decomposition was employed in the previous chapter to obtain a load balance for a regular 2D computational domain structure. But many computational domains are not regular, and an alternative method of load balancing must be employed to effectively exploit the multicomputer's computational potential.

Simulated annealing is one algorithm that has been successfully used to load balance static domains. A brief description of simulated annealing is given as a background on load balancing techniques for static domains. Two other techniques, the Orthogonal Recursive Bisection (ORB) ([Fox88b]) and Eigenvector Recursive Bisection (ERB) ([Pot89], [Bar82], and [Bop87]), both analyzed and compared by Williams [Wil90], are described for static computational domains. For dynamic computational domains, the load balancing technique of Brugè and Fornili [Bru90] is reviewed. Finally, systolic compression expansion (SCE) load balancing is introduced as a means for realizing a continuous real-time load balance condition.

7.1 A Partial Taxonomy of Load Balancing[65]

Computational problem domains for MIMD-class simulations are organized into 4 categories. These are listed in Table 7.1. Each domain is tabulated with a brief

---

[65] The work of R.D. Williams [Wil90], of the CalTech Concurrent Computation Program is used as a basis for this taxonomy.

description of the domain's load balance and data decomposition requirements.

The regular domain, as defined here, encompasses computational problems restricted to highly symmetric structures, where the geometry remains fixed during simulation execution, and inspection can be used to obtain a load balance.

The second class discussed here is the static computational domain. The geometry assumes a more complicated, less regular form or structure; load balancing via inspection is not possible. An algorithmic process is needed to compute a load balance prior to running the simulation.

The dynamic domain involves simulation over either regular or static geometries, but the data elements exhibit a temporal or spatial dependence which physically causes their relocation to other processors during the simulation's execution. The data elements are in constant motion between processors, and this attribute imposes a strong requirement for adjusting the global processor load balance to maintain a margin of multicomputer efficiency. The frequency of load balancing operations is not subject to hard timing constraints.

Table 7.1 Computational domains and their load balance characteristics (after Williams [Wil90]).

MIMD-class Problem Domain	Load Balancing Characteristic
Regular	Simple symmetrical computational domain structures; usually rectangular; decomposed by inspection.
Static	Irregular (non-symmetrical) computational domain geometries; decomposed and balanced via algorithmic processes prior to simulation execution.
Dynamic	Irregular or regular computational domain structures; continuous, but an asynchronous and/or lazy balance operation is necessary during simulation execution to maintain minimum dead-time and eliminate sequential bottlenecks.
Real-time	Irregular or regular computational domain structures; continuous balancing under hard real-time constraints necessary; the balance contribution becomes an intimate part of the simulation's fidelity.

However, real-time load balancing must obey all predictability and hard-timing constraints that are imposed on the simulation. The continuous readjustment of the processor load and data decomposition becomes a tightly coupled and intimate partner with the primary simulation.

7.1.1 Regular Domains

The first category is the regular domain; we have already encountered this problem in the previous chapter. The load balancing characteristic in Table 7.1 explains that the computational domains within this class are symmetric, and can be partitioned by inspection. A regular domain possesses a fixed and motionless internal geometrical structure, and the boundary conditions are well understood and finite; however, the solution may evolve with time, such as during the propagation of a waveform along a guide. Boundary value problems that have spatially and temporally decoupled eigenfunctions typify this class, including the classical partial differential equations of physics and mathematics [Arf70].

Critical industrial processes often depend on solutions to these classical equations, since regular geometries substitute as approximations for intricate ones. Although a first-order solution may not provide any indication beyond an order-of-magnitude in terms of cost, material, labor, or physical efficiency, that an answer can be obtained which is "reasonable" is usually justification enough for some business purposes. Therefore, it is important to recognize the role of the classical partial differential equations for their surrogate utility in simulation.

The numerical example of the previous chapter (section 6.7) illustrated how one can proceed to perform a data decomposition which produces a computational load balance. Achieving a computational load balance is a principal requirement and objective of disciplined multicomputer software engineering. The load balance operation adds additional complication to the design effort for multicomputer simulations. However, regular domains which are easily decomposed provide an excellent starting point for approaching more complicated situations. Patience must be exercised while developing a data decomposition for even the regular domain.

7.1.2 Static Domains

Static computational domains encompass a larger group of regular problems; they include geometrical structures which cannot be easily decomposed by inspection. If the inspection or "eye-balling" activity does not or cannot rapidly produce a load balance, an algorithmic process must be employed. A static computational domain remains fixed during the simulation; the geometry remains frozen, and communication is only necessary to complete computations dependent on data elements that share common boundaries. However, the data decomposition often becomes non-trivial, owing to the irregularity of the computational domain structure. The eye is a terrific pattern recognizer, but load balancing a static domain over the physical topology of an n-dimensional hypercube is too complicated for our biological computers to perform effectively. An algorithmic solution is needed.

Simulated annealing is a powerful technique for static domain load balancing which is outlined in section 7.2. An additional technique, called Orthogonal Recursive Bisection (ORB) is simple to implement, although it is primarily applicable to domains with orthogonal boundary structures like a rectangular plate or parallelepiped. ORB may be used when the internal data element representation of the computational domain is also non-symmetric, like the structures found in an integrated circuit.

Eigenvector Recursive Bisection (ERB) is an alternative method. It is more complicated in that eigenvectors must be computed from the adjacency matrix of the computational domain. Both are briefly mentioned in this taxonomy.

The load balance operation for a static domain is performed just once, although the algorithmic load balancing process may require several minutes or more to complete. With simulated annealing, optimization is performed on a cost-function over a configuration space of processors, data elements, and factors involving communication and computation time for the multicomputer host. The ORB load balancing algorithm does not consider communication factors. This algorithm achieves data decomposition through geometric construction techniques -- deconstruction -- of the computational domain. The ERB load balancing algorithm incorporates communication factors on a functional basis. A load balance and data decomposition is achieved through the eigenvector solution of a sparse matrix. The eigenvectors located with this algorithm are derived in relationship to the computational domain's connectivity: the data dependencies between elements as represented by communication between them.

One can employ simulated annealing with a *scattered decomposition* [Mor85]. A scattered decomposition involves the arbitrary distribution and assignment of data elements, such as a finite element mesh, or collection of data structures, throughout the multicomputer platform. The data element assignment is known *a priori*, and this configuration is evaluated with respect to the cost of communication and computation. By randomly moving data elements between processors under the guidance of a simulated annealing algorithm, a load balance is generated.

Configuration optimization of a scattered decomposition via simulated annealing can produce a *local minimum*. A local minimum, of which several are likely to occur for any large ensemble of processors and data elements, are inappropriate load balance candidates. They represent an optimal value that is relevant over a small region of the simulation's data decomposition and the multicomputer's processor configuration. Local minima can be avoided by using an annealing schedule. This facet of simulated annealing is discussed in section 7.2.

The two recursive bisection algorithms operate on a different premise. The ORB method of Fox attempts to organize a load balance by partitioning the geometrical domain through repeated subdivision. With each subdivision, more computational domains are formed, isolating small clusters of elements into islands. Each additional island formed through the bisection process creates more internal surface area, along which communication between the newly formed island and its neighbors becomes necessary. Eventually, each bisection generates islands with roughly the same quantity of elements. ORB is simple to implement elegant because of this simplicity, but it is limited to orthogonal domain structures.

The ERB method is not restricted to orthogonal domain structure. The location of eigenvectors, the principle axes of symmetry for a domain, affords a powerful method for data decomposition in static domains. ERB constructs a matrix through the discretization of a cost function operates which reflects communication and computation requirements of a configuration confined to some geometric enclosure. Each eigenvector solution to the cost-function matrix representation is used to partition the adjacency matrix arising from the communication dependencies of the data elements. The following function is used describe load balance cost. It is actually derived from the Hopfield model of neural computation, and the end result appears as [Wil90]:

$$(-D + A + \xi I - \mu^{-1}E)\vec{u} = N\vec{u}$$

where D is a diagonal matrix whose elements are the degrees of the graph nodes, I is the identity matrix, and E is the matrix with 1 *in eachentry. The matrix N is symmetric. The value of* μ *corresponds to the importance of communication.* ξ *is an arbitrary constant.* 7.1

The adjacency matrix A of the graph expresses the connectivity between the data elements located within the computational domain. The eigenvectors, $\vec{u}$, of this matrix are used to bisect the domain and obtain a decomposition with minimum communication overhead, as represented by the factor μ.

7.1.3 Dynamic Domains

The dynamic domain is characterized by the aperiodic or asynchronous redistribution of data elements during the course of simulation. The workload on each processor is adjusted through the redistribution of data elements, or the migration of processes. An executive-level protocol communicates processor workload conditions via a broadcast or multicast that signals under-utilized or over-utilized processors. Strategies of this sort are used in many distributed environments, like the telephone system which must redirect calls and transactions through alternate routing exchanges when a failure or overload condition arises.

An unbalanced multicomputer simulation executing with synchronous message-passing conditions will exhibit poor efficiency, owing to the accumulation of *dead-time*. All processes will remain in a blocked state until the most heavily-laden node reaches the communication phase, where a sequential bottleneck exists. The sequential bottleneck generates dead-time and enforces unwanted and unnecessary global synchronization. Redistribution of the workload is needed to minimize the impact of dead-time on multicomputer efficiency.

The redistribution of work during a dynamic simulation produces a time-averaged load balance, $B_l(\mathrm{t})$, which is defined as:

$$B_l(t) = \int_{\tau_i}^{\tau_f} B(t)\,dt,\ \tau_f > \tau_i$$

$$\textit{where } B(t) = \sum_{\substack{\text{Processors}\\ i=0}}^{N} B_i(t)$$ 7.2

and τ_i, τ_f *are the sample interval end points.*

Each $B_i(\mathrm{t})$ represents the instantaneous workload on processor i, in terms of data elements -- bytes -- completed each second. $B_l(\mathrm{t})$ gives an aggregate measurement of the memory bandwidth associated with each processor's load. "An important measure of the performance of an algorithm for dynamic load-balancing is the number of elements migrated, as a proportion of the total elements." [Wil90] If processor A has a workload twice that of a companion processor ß, $B_\alpha(t) = 2B_\beta(t)$. Similarly, if processor γ has 8-times the load of another, γ has ⅛th the throughput of a nominally loaded one. A nominally loaded processor is termed a *standard processor*, and can be used to obtain an absolute load balance reference. Standard processors execute with nominal computational bandwidth, and can be calibrated from a sequential iteration of the problem prior to running the multicomputer version.

The time required for the standard processor to complete a quantum of work, such as one iteration of an algorithm, or the time to move a fixed number of bytes from one memory location to another, can be easily measured in a sequential system that is constructed with the same processor host and clock speed used by the multicomputer platform. The standard processor load is denoted by the symbol B_s.

The time-averaged load balance $B_l(t)$ is calculated by adding the contributions of $B_i(t)$ within the interval $\{\tau_i, \tau_f\}$. This procedure yields, for each interval $\{\tau_i, \tau_f\}$, the following value for $B_l(t)$:

$$B_l(t) = (\tau_f - \tau_i) * \sum_{\substack{processors \\ i=0}}^{N} B_i(t) \quad (Bytes) \tag{7.3}$$

where the integral over time has been replaced by the interval $(\tau_f - \tau_i)$.

The 8-bit byte is used to describe load balance measurement. Although in many instances, a fixed data structure will be used to represent the fundamental unit of message-passing between processors, and this quantum should be substituted, since more meaning is conveyed by a quantity like events or elements, rather than bytes. With N multicomputer nodes, the node average load balance is simply given as:

$$B_N = \frac{B_l(t)}{N} \tag{7.4}$$

Sampling the multicomputer's instantaneous load balance condition each $\Delta t = (\tau_f - \tau_i)$ seconds generates a stream of data suitable for simple statistical computations, such as the mean and standard deviation of load balance. All processors must have their load balance parameters examined within the same Δt time-frame. When continuously displayed on a CRT, or printed on an oscillograph, a real-time history of the load balance profile is obtained that gives quantitative experimental information on the overall multicomputer efficiency and utilization by the simulation and the accompanying load balance strategy (Figure 7.1). The relative workload per processor, expressed as a percentage derived from the quotient, is $B_i(t)/B_s$. This value provides visibility into individual processor performance, and can help to isolate sequential bottlenecks when visualized.

By reporting and accumulating this information, a small sacrifice of message-passing bandwidth is made. Trading-off part of the message-passing bandwidth during the initial stages of tuning a multicomputer simulation for load balance monitoring purposes is considered wise and prudent. In dynamic computational domains, the simulation executes with a fixed load balance configuration until the simulation executive detects or determines that a re-balance is needed. Re-balancing the simulation in this lazy fashion may reduce total wall-clock time, assuming that load balancing accounts for a fraction of the simulation's purpose.

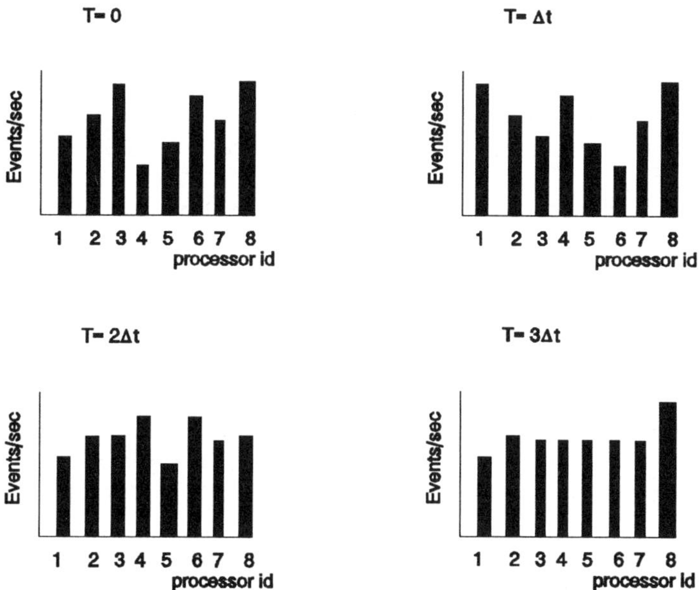

Figure 7.1 Time-dependent load balance profiles for an 8 processor multicomputer. (Byte Magazine, June 1991 © McGraw-Hill, Inc., New York. All rights reserved.)

A classical problem which illustrates the necessity for a dynamic multicomputer load balance strategy is the WaTor simulated biological population game devised by A.K. Dewdney [Dew84]. The game grid is populated by idealized sharks and fish that can breed, live, and move randomly in an two-dimensional simulated ocean. Chapter 17 of Fox *et al.* ([Fox88a]) gives a detailed discussion of this game and its effectiveness for illustrating dynamic load balancing strategies. Similar load balancing strategies are needed in multicomputer solutions to the n-body problems encountered in molecular dynamics or quantum chromodynamics. Two techniques for dynamic load balancing, as applied to molecular dynamics problems, are described in section 7.3.

The dynamic and the real-time domains employ unique load balance strategies to achieve simulation objectives. A dynamic load balance arises by the institution of a global determination mechanism that initiates balancing operations as necessary, but without regularity or predictability under hard real-time constraints. A real-time load balance must satisfy these requirements. The real-time simulation and its attendant load balance operation are inseparable.

7.1.4 Real-time Domains

Real-time multicomputer simulations represent the forefront of software engineering, although research in this area has not widely penetrated many popular professional publications.[1] Image processing is one of the most popular, plausible, and primary applications for real-time multicomputer simulations ([Jun90] and [Chu91]). Image recognition by computer is one of the Grand Challenges and goals of the High Performance Computing Program. As outlined in Chapter 4 on real-time systems, a real-time computation is characterized by temporal and algorithmic predictability. The load balance element of a real-time multicomputer simulation must also obey these criteria.

A load balanced multicomputer simulation of a real-time domain implies that a continuous readjustment and assessment of the multicomputer workload is predictably executed and maintained. This condition differs markedly from static domains, or even the dynamic domain, where continuous load balancing is not a synchronous or predictable necessity. Suppose, for instance, that a multicomputer simulation running on N nodes must maintain a bounded global dead-time,[2] T_d, such that the frame-time, $T_f + T_d < D$ seconds. The frame-time for each iteration or simulation step, T_f, may consist of computations and communications, as necessary, to achieve fidelity goals and modeling requirements. T_d accounts for the global differential between packet sizes and communication requirements such that all message-passing should complete during T_f, but because of network-routing collisions, unequal packet lengths, or other factors, fails to do so.

If the simulation encounters a load balance disparity which instantaneously upsets this condition for one frame on one processor, then the entire frame becomes invalid for the multicomputer. The analogy between the sequential real-time simulation on one processor, or the shared-memory multi-processor environment is accurate: a deadline must be met by the simulation to validate a result, computation, input, or other operation.

In the sequential environment, if one process misses a deadline, the violation of a predictability constraint invalidates the result. The same is true in the multicomputer system. The restoration of a load balance must then be accomplished in some manner to preclude other frame invalidations, and perhaps avoid safety mishaps. The load balance difficulty is slightly ameliorated if the simulation executes a data parallel computation in the single-program multiple-data (SPMD) sense. In this case, each processor runs the same computation, and only the dataset size changes

---

[1] See the Suggested Reading section at the end of this chapter.

[2] A discussion of global time synchronization is given in Chapter 8. Obviously, the notion of real-time load balancing is tightly coupled to the concept of global time regulation and maintenance.

between processors. To prevent frame over-runs, a processor must not obtain a share of the dataset that forces a violation of the bounded global dead-time requirement.

This implies that all message passing transactions, assuming point-to-point nearest-neighbor communication, must conclude within the global dead-time limit of D seconds. The aggregate packet volume must not exceed size $L_{max} \leq (T_f + T_d + t_l)/t_c$, where t_c is the time to communicate one byte, and t_l is the message latency[3] as described in section 6.3.2. If $t_c = 1 \times 10^{-7}$ s/byte, t_l is 1×10^{-5} seconds, and T_f is 20 ms and t_d is 13.333 ms, this gives a maximum packet size $L_{max} = 3.33 \times 10^5$ bytes, or a little less than the typical disk burst transfer on a personal computer.

A total network bandwidth of .3 Gbytes/s is required to support a real-time multicomputer simulation with 1000 simultaneously communicating point-to-point process peers in a packet-switched routing environment, running on 1000 separate processors. The packet size constraint, L_{max}, for point-to-point process peer communication is well within the limit of current silicon hosts, like the transputer, Intel iWarp cell, or Intel iPSC-860 node. So long as the aggregate packet volume does not exceed L_{max}, at least the global dead-time limit will not be exceeded.

Alternatively, a control parallel simulation executing under similar constraints must also obey the same restriction, but isolating the culprit which causes the frame overrun is much harder. The control parallel simulation may contain many unique processes, and each may independently contribute to the packet volume. Locating the conditions that force an overrun can be exceedingly difficult. The culprit process may only violate global dead-time constraints under peculiar simulation states and temporal conditions that are difficult or impossible to artificially orchestrate, initialize, or simulate. The application of SAR methodology is no help in this instance. While the simulation's logical structure is preserved with SAR kernel congruency, the simulation's temporal characteristics are not. Temporal consistency can only be assured within the multicomputer.

To ensure temporal consistency under real-time load balance constraints, the systolic compression expansion (SCE) load balancing technique is described in section 7.4. This method is a representative candidate for real-time load balance operations. Discussion of this method is supplied within the context of real-time graphics rendering as posited by the scalable concurrent visualization system of Chapter 2.

7.2 Simulated Annealing and Static Techniques

The simulated annealing approach is usually employed when a data decomposition by inspection is not possible in a static domain. *Simulated annealing* [Kir83], as applied to load balancing, performs optimization on a cost-function that expresses the expense of computation and communication in the physical multicomputer system. To realize optimal multicomputer efficiency, a compromise between the independent computation and communication aspects of the simulation must be implemented. Simulated annealing load balance strategies perform optimization on the data decomposition

---

[3] Latencies approaching a few μsec are possible with the Intel iWarp cell and the hotly anticipated T9000 transputer from Inmos.

configuration. The load balance for a static domain is often carried out before the primary simulation body executes, usually during an initialization phase.

Simulated annealing, as its name implies, algorithmically simulates the thermodynamic process of annealing. In a multicomputer environment, the number of data elements, their computation, and communication requirements are annealed within a discrete configuration space spanned by the processor nodes and the data element communication length requirements, to minimize packet message-passing time, T_{ML}, the time consumed between packet transmission from a source to a destination.

Descriptions of the simulated annealing algorithm, and how it relates to thermodynamics, are often given for physical purposes. Many descriptions in the literature are known; the one given by Press *et al.* ([Pre86])[4] is particularly eloquent:

> "At the heart of the method of simulated annealing is an analogy with thermodynamics, specifically with the way that liquids freeze and crystallize, or metals cool and anneal. At high temperatures, the molecules of a liquid move freely with respect to one another. If the liquid is cooled slowly, thermal mobility is lost. The atoms are often able to line themselves up and form a pure crystal that is completely ordered over a distance up to billions of times the size of an individual atom in all directions. This crystal is the state of minimum energy for this system. The amazing fact is that, for slowly cooled systems, nature is able to find this minimum energy state. In fact, if a liquid metal is cooled quickly or 'quenched,' it does not reach this state but rather ends up in a polycrystalline or amorphous state having somewhat higher energy.
>
> "So the essence of the process is *slow* cooling, allowing ample time for redistribution of the atoms as they lose mobility. This is the technical definition of annealing, and it is essential for ensuring that a low energy state will be achieved."

Slow cooling is accomplished via an annealing schedule. The schedule specifies the rate at which the configuration temperature is lowered to obtain a globally optimal multicomputer load balance. This schedule is applied to the evaluation of the cost-function after the trial or random relocation of N elements.

The cost-function used by Williams [Wil90] for simulated annealing load balancing operations is derived via a perturbation method. The first component of this function, termed H_{calc}, is a Hamiltonian representation of the "energy" describing a multicomputer computation based on N data elements conducted over P processors when each is equally loaded. The communication time between elements, the second component, termed H_{comm}, is to be minimized.

All calculations conducted for a data element -- the "number crunching" component of the simulation -- is accomplished with a fixed set of arithmetic

---

[4] Numerical Recipes, The Art of Scientific Computing, Cambridge University Press, Copyright © 1986, New York. All rights reserved.

operations. Microcoded instructions are optimal in that they execute as fast as the silicon permits -- cycle times and instruction lengths are preordained by the silicon. However, we can optimize the communication aspect, since the physical locality of data elements relative to each other in the multicomputer can be arranged to minimize T_{ML}. A proportionality exists between the per-element computation time, and the per-element communication time. If the per-element computation requirement is large, as measured in floating-point operations, then communication accounts for a relatively small amount of work. Therefore, the cost-function H assumes the form [Wil90]:

$$H = H_{calc} + \mu H_{comm} \qquad 7.5$$

where H_{comm} represents the communication cost contribution to the Hamiltonian, and μ is a proportionality constant that accounts for the ratio of computation to communication. μ may be represented as [Wil90]:

$$\mu = \frac{\#\ of\ floats\ communicated}{\#\ of\ floats\ multiplied\ or\ added\ per\ data\ element} \qquad 7.6$$

The cost-function is finally determined to be:

$$H = \frac{P^2}{N^2}\sum_q N_q^2 + \mu\left(\frac{P}{N}\right)^{\frac{(d-1)}{d}} \sum_{e \leftrightarrow f} 1 - \delta_{p(e),\, p(f)} \qquad 7.7$$

where P represents the number of processors, N the number of data elements, μ the communication to computation ratio, and the functions $p(e), p(f)$ denote the processor assignment of the data elements.

The first term, H_{calc}, accounts for the computation portion of the simulation. The summation is over all data elements q in the decomposition, on a per-processor assignment basis. The constant term, P^2/N^2, reflects a normalization scale factor to keep the cost-function contribution to near unit value for all data elements q.

The second summation, over $e \leftrightarrow f$, for the H_{comm} contribution, in the second term of the cost-function, expresses the communication cost between two data elements e and f. The quantity $(P/N)^{(d-1)/d}$ accounts for the dimensionality of multicomputer. Note that the communication portion of the cost-function yields a zero value when two data elements reside on the same node, owing to the orthogonality property of the Kronecker delta.[5]

A strategy to guide the migration of data elements between processor nodes is needed which is purely random and bias free. One such strategy is known as the Metropolis method [Met53], which "can be generate trial decompositions (using random numbers -- a Monte Carlo) such that the probability of a particular decomposition occurring is proportional to $e^{-S/T}$ [Flo86]."

---

[5] $\delta_{i,j} = 1$, when $i = j$, and $\delta_{i,j} = 0$, when $i \neq j$.

"T is analogous to temperature. If the simulation is begun at some initial value of T which is large ("hot"), and then T is slowly lowered towards 0 (annealing), the system will be driven towards the configuration of minimum S [Flo86]."[6]

An approximate pseudocode representation of the Metropolis method, combined with a very simple annealing schedule, is shown below. The algorithm is based on the work of Flower *et al.* ([Flo86]). The annealing schedule slowly lowers the temperature of the configuration, and some experimentation may be necessary to determine when δT should be subtracted from the configuration temperature. It may be appropriate to reduce the configuration temperature when a fraction N_q/N data elements have been successfully relocated to new processors. This will be evident by a consistent lowering of the cost-function. However, the temperature correction quantity, δT, must also be chosen carefully. If δT is too large, the temperature will drop too quickly, and the configuration will not reach an annealed condition (a pure crystalline state), or globally minimum state, but will instead reach a local minimum (a polycrystalline state). Careful choice of the annealing schedule parameters is necessary to prevent the configuration from rapidly quenching, and settling into an undesired local minimum. Guidelines and heuristics have been developed by Hajek [Haj88] for selecting annealing schedule parameters.

Evaluate $H_{old} = H$ *with an initial scattered data decomposition*
$T = T_0$.
$\delta T = .999$
for each configuration temperature $T > 0$
  for each data element i
  {
    move the element$_i$ *to a new processor; compute* H.
    compute $\Delta H = H - H_{old}$. 7.8
    if $\Delta H < 0$, *accept the move.*
    if $\Delta H > 0$, *accept the move with probability* $e^{-\frac{\Delta H}{T}}$.
    if the move is rejected, return the element to its original place.
    $H_{old} = H$.
  }
  $T \mathrel{*}= \delta T$

Flower *et al.* [Flo86] reported on simulated annealing load balancing as applied to four unique initial data decompositions: (1) a "Random Start" condition, where the data elements were randomly assigned to processors; (2) a "Node Zero" condition, where all data elements were initially confined to a single processor node; (3) a "List

[6] Flower, *etal.* use the symbol S for the cost-function in place of the H used by Williams.

Start" condition, where the data elements were assigned to nodes based on an initial value associated with the data elements, such as strain energy or stress; (4) a "Grid Start" where the geometry was enclosed in a rectangular box, and simply divided by inspection, with data element assignments given by region. Each initial data decomposition was subjected to simulated annealing and a data element relocation strategy. Flower *et al.* concludes that any of the four configurations requires approximately the same amount of time for the configuration to reach an annealed state. This study presents a parametric assessment of the communication to computation ratio, and how this parameter affects the configuration energy H, and the annealing schedule.

7.3 Dynamic Technique

Dynamic multicomputer simulations consist of mobile data elements, such as atoms, molecules, or particles, which are subjected to kinetic laws and equations of motion. At each iteration of the simulation, particles acquire energy through repeated interaction with others in the presence of electrical, gravitational, atomic, or nuclear forces. As a result of these interactions, the particles undergo physical displacement to a new spatial coordinate (x,y,z) as a function of time. In classical systems, the particles obey Newtonian equations of motion. In the quantum domain, the time-dependent Schrödinger equation governs electron and nuclear motion. N-body problems are extraordinarily important for many scientific endeavors. Scientific disciplines such as molecular dynamics, pharmaceutical design, particle physics, and astrophysics rely on theoretical analysis and simulation by employing n-body systems of equations to model physical phenomena.

Interacting particle systems in conservative force fields are termed *n-body* problems; each particle or body in the system interacts with all of the peers, except (usually) itself. In general, a system of n interacting particles requires n(n-1)/2 evaluations to derive a new potential energy, or positional configuration for the ensemble. The $O(n^2)$ computational metric implies that solution times grow in proportion to the square of the number of interacting bodies.

This growth characteristic places restrictions on the size of the problem domain which can be economically solved. The solution to problems with large numbers of particles calls for substantial computer resource. The solution to idealized n-body systems with small numbers of particles are useful for exploring many physical theories. However, the solution to problems that confound mankind are usually not ideal, and have extraordinary characteristics and particle populations. A multicomputer platform can be organized into an effective edifice for n-body simulation.

In a classical n-body problem, the particles constantly move under the influence of a force field generated by the potential energy of attractive or repulsive forces. At each instant of time, the coordinates of each particle change. When the simulation is initialized, a starting configuration is constructed where particles are assigned to specific processors.

Each processor is allocated a portion of the computational domain geometry. When the particles move in space, their coordinates change, and they appear within the sphere of computational influence of a new processor. Each particle requires the same amount of computation, so the addition of extra particles adds to the processor's computational burden. A technique for adjusting the computational burden must be employed, or a serious sequential bottleneck will emerge.

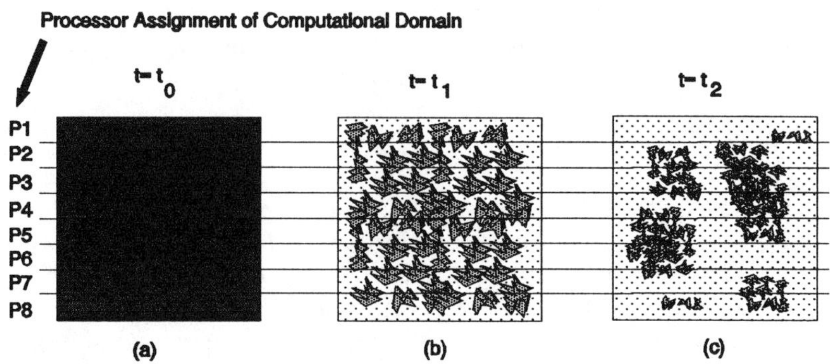

Figure 7.2. The evolution of the load imbalance in a dynamic simulation (after Brugè and Fornili [Bru90]).

Figure 7.2 illustrates the problem of dynamical load balancing without readjusting the computational domain responsibilities of the multicomputer. Figure 7.2a shows an initial domain decomposition and processor assignment at $t= t_0$. At time $t= t_1$ in Figure 7.2b, the particle system begins to demonstrate clustering owing to the time-evolution of the physical potential and forces. The formation of particle islands, while indicative of the phenomena under study, shifts the load balance toward an inefficient value. Further time-evolution of the particle system produces larger clusters, with highly unequal particle populations evident between the processors. Figure 7.2c shows the system at $t= t_2$.

The novel load balancing technique of Brugè and Fornili is ideally suited to counteract the conditions shown in Figure 7.2. Their method incorporates a feature to adjust the computational domain responsibilities of each processor in response to particle motion. The technique is actually quite simple to implement and very effective. They employ a toroidal multicomputer topology for their molecular dynamics simulation.

Their load balance method operates through the detection of particle migration between processors. Since the particle count changes in response to migration resulting from molecular forces, the algorithm contains a particle count parameter which activates the load balance function when the count on any processor changes by an amount Δ C. The load balance method proposes new spatial domain limits for a processor, and evicts any particles which are determined to lie outside the new domain limits. The evicted particles arrive at neighboring processors, which also execute the load balance strategy independently.

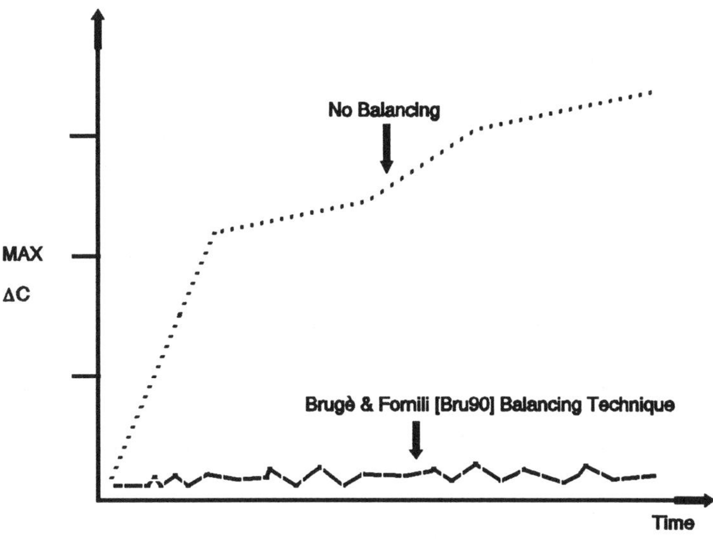

Figure 7.3. Qualitative behavior of Brugè and Fornili load balancing. The maximum particle count difference between two cpus is Δ C.

The spatial domain limits for the top or bottom of each processor's domain can independently expand or contract by a preset value of δ B units. The processor's domain limits are checked to verify that the top or bottom of a proposed domain limit shift expansion or contraction by δ B units does not protrude into the bottom or top of a neighboring domain. The domain shift process is governed by two functions: one which "proposes" new domain limits, and a comparison function arbitrator which actually adjusts the limits by consulting with domain limit propositions supplied by nearest neighbor processors. This control loop maintains a nearly constant processor load during the simulation. The qualitative behavior of this load balancing technique is shown in Figure 7.3, expressed as the maximum particle population difference on all processors as a function of time Δ C.

Computational domain boundary adjustment techniques supply a simple and effective method for dynamic multicomputer simulation load balancing. The n-body simulations with large particle counts are generally not concerned with continuous domain boundary adjustment, but only after the imbalance reaches a threshold. The threshold tolerance remains fixed during the simulation. A dynamically adjusted threshold tolerance may lead to a load balance instability -- a load imbalance resonance -- if the adjustment frequency is not closely coupled to the load balance operation itself.

The luxurious and lazy frequency of load balance operations in the dynamic domain is inconsistent with a hard real-time multicomputer simulation, where each simulation frame serves an essential role in a tightly orchestrated and accountable structure. A real-time simulation structure includes a load balance operation, along with the algorithmic representation of the problem domain. During each frame, the global load balance condition must be assessed and measured to ensure and satisfy predictability goals and maintain safe operations. The load balance becomes temporally significant and highly visible owing to the simulation's timing constraints. The real-time load balance trigger resides with the temporal rhythm of simulation.

7.4 Real-time Technique

A hard real-time data parallel multicomputer simulation must obey all of the analogous predictability criteria established for sequential real-time simulations. Data parallelism, as embodied via the single-program multiple data (SPMD) formulation, permits the organization of a replicated logically concurrent process structure, where the identical computation and communication sequence occurs throughout the multicomputer platform. Replicated logical concurrency provides a true advantage for debugging the computational and message-passing simulation components, since SAR methodology guarantees congruency between the logical and physical implementations. Unfortunately, the timing and predictability elements have no substitute or surrogate debugging methodology which is equal in power and capability.[7]

As a basis for discussion, the scalable concurrent visualization system discussed in Chapter 2 supplies a nice hypothetical framework on the real-time load balance issue. Recall that the SCVS (see [Ste92c]) visualizes datasets in a highly data parallel fashion, since each node in the multicomputer is equipped with a small framebuffer and display device. The synchronization issue for this multicomputer simulation, while substantial and non-trivial, is pushed aside for now, and discussion is reserved until Chapter 8. The load balance problem is closely related to the synchronization problem, but a qualitative treatment can be given. Figure 7.4 illustrates the load balance anomaly for the SCVS tessellated display architecture.

The load balance problem arises from the continuous transformation of the geometry in response to user interface commands, or the automated solution to an equation of motion. Since the geometry is represented as discrete display primitives like polygons, splines, patches, or NURBS[8], the load balance for each processor of the display subsystem changes when the display primitives migrate to different processors. What method or methods are available to continuously ameliorate the

---

[7] Perhaps the parallel random access machines currently on the computer architect's drawing board will provide an avenue to develop a congruent real-time stand-alone router (RSAR) kernel.

[8] Non-uniform rational B-splines.

ensuing load imbalance?

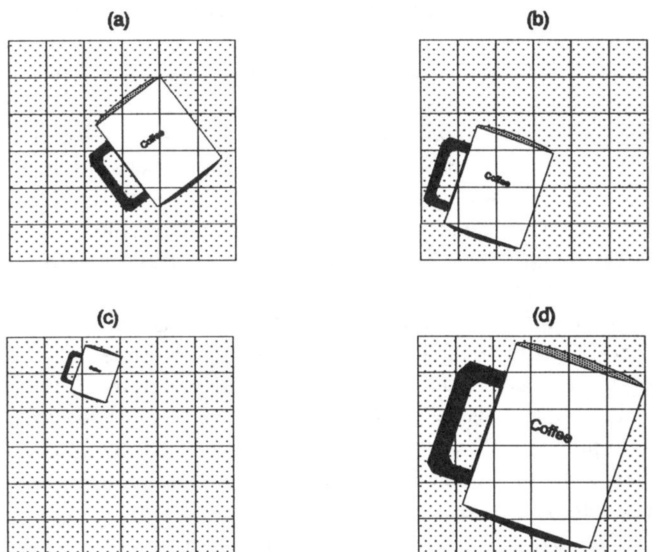

Figure 7.4 The load balance anomaly for tessellated displays. Note how each tessellation element acquires a greater or lesser portion of the display primitive set during transformation.

The available literature on MIMD-class tessellated display architectures is non-existent.[9] But the crux of the load balancing problem is relatively simple: For each simulation frame, T_f, determine which node carries the maximum quantity of display primitives, B_{max}. Expedient determination of the most laden node yields the maximum simulation dead-time, relative to the least loaded node, since all processors must wait for the slowest to complete computations and enter the message-passing component of the simulation. If the global dead-time is known during each frame, the display update of all tessellation elements can be adjusted and managed to occur simultaneously, so as to avoid display *twinkle*. The global dead-time, T_d, in the SCVS case is defined as:

$$T_d = |\max\{ B_i(t) \} - \min\{ B_j(t) \}| * \frac{t_{calc}}{t_{vertex}}, \ \forall\ i,j \in SCVS\ nodes. \qquad 7.9$$

Where t_{calc} is the flop time, and t_{vertex} is the vertex processing time.

If an asynchronous update of each tessellation element dominates, the output will not

---

[9] At least to the author's knowledge as of Jan. 1992.

be coherent; one display element will perform framebuffer output out-of-phase with the other tessellation elements. The twinkle effect may lead to nausea in the observer, because the visual clues will appear visible in a random sequence throughout the tessellation, and possibly confuse the senses.

Fortunately, the temporal aspect of human visual sense is not especially acute [Wol83]. The eye's threshold detection between stop action and continuous motion hovers around 25 times each second; this physiological limit establishes the dead-time constraint. The twinkle effect can be minimized if each tessellation is updated within about 40 ms of a global reference. Therefore, so far as the SCVS load balance condition is concerned, knowledge of the most heavily laden node supplies a trigger to release framebuffer output. Additionally, the 25 Hz sensorial threshold constrains the quantity of computation and communication which can be conducted, and still achieve the appearance of continuous motion output at the tessellation.

In Chapter 2, the technical discussion concluded that approximately 110 floating-point operations are necessary to perform the geometric transformation and graphical rendering operations for each vertex (x,y,z,n). Assuming that each FLOP consumes .2 μs implies that 22 μs/vertex elapse. Approximately 1820 vertices can be transformed over a 40 ms interval with this FLOP time. In contrast, L_{max}, the aggregate packet volume between nearest neighbors in this time interval equals 1.66 x 10^4 vertices (4 x 10^5 bytes ÷ 24 bytes/vertex). The message-passing component has the upper hand in this application; the communication bandwidth is far in excess of what the cpu can tolerate. A 9:1 ratio of communication to computation is found. Nine-times as many display primitives of fixed representation can be passed than processed. This ratio is disturbing from the computational point of view: more processing power, such as a dedicated graphics processing unit, will substantially lessen this ratio.

The 9:1 ratio portends consequences for the maximum sustainable continuous balanced rate of SCVS operation. Since more data can be communicated per unit time than can be expeditiously processed, message-passing operations must be restricted to hold down the volume. Placing a constraint on the message-passing implies that the continuous transformation of the geometry slows as well. Specifically, the 9:1 ratio implies that only 1/9th of the available message-passing bandwidth can be utilized at T_f = 40 ms. Therefore, realizing the 25 Hz simulation frame rate, Ω_f, implies that $L_{max} \approx$ 40Kbytes can be exchanged between processors and achieve our fidelity objective.

Armed with this knowledge, one can proceed to organize a data decomposition over the viewing frustum, such that the number of frustumettes created by decomposing the field of view contain equal volumes. The goal here is to structure the aggregate message-passing to permit a processor to either gain or lose less than L_{max} during each interval T_f. The viewing frustum decomposition can be easily accomplished via plane geometric methods, and the algorithm for performing a decomposition within a K by M tessellation is simple to construct[10] once the global parameters defining the viewing frustum are known (e.g., field-of-view, aspect ratio, and the near and far clipping planes). Figure 7.5 illustrates a perspective rendering of the decomposed viewing frustum.

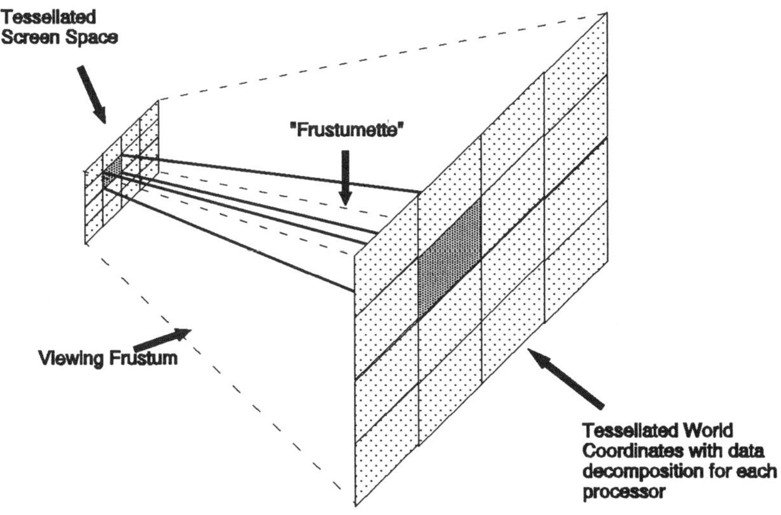

Figure 7.5 Viewing frustum decomposition for the SCVS tessellated display.

The volume of each frustumette is exactly equal, and can be proven as follows. Assume that the decomposition of the original viewing frustum subtends a field of view of Θ degrees. By placing a squard grid of N by N equally spaced points at both the near and far clipping planes of the original frustum, and connecting parallel lines between the gridpoints on the near and far planes, frustumettes are formed where each subtends a field of view $= \Theta / N$.

The near and far clipping planes subdivided by the N by N equally-spaced gridpoints form quadrilateral areas, called A_n and A_f. Each frustumette that is formed in this way encloses the same volume of the original frustum as shown by Equation 10, but subtends a different portion of the field of view. Each frustumette is therefore responsible for displaying that portion of the decomposed dataset which resides within its field of view.

---

[10] The author assumes a static dataset. A robotics or time-evolutionary dataset complicates the load balance insofar as the time-dependent component of the simulation may create more than L_{max} bytes per processor per frame.

$$V_f = \int_V d(vol) = \int_0^z A(z)\,dz, \text{ where } A(z) = A_n + \left(\frac{(A_f - A_n)}{D}\right) z\,dz, \text{ where}$$
$$D = D_f - D_n \text{ is the distance between the far and near clipping planes.} \quad 7.10$$
$$\therefore V_f = \int_0^D \frac{1}{D}[A_n * D + (A_f - A_n) * z]dz = (A_n + A_f)\frac{D}{2}$$

During each simulation frame, a new position vector arrives at each node from the user interface or an equation of motion, and the geometry will undergo transformation in response. During this process, a portion of each node's (frustumette's) display primitives may be clipped out of the field of view. Prior to rendering this geometry, the data which lies outside a frustumette's field of view, but which is still node-resident, must be communicated to a neighboring frustumette's node for processing. Likewise, the node which sent a portion of its graphical display primitives may receive an in-bound packet from a neighbor. The in-bound and out-bound packets may not be of equal size, since the geometry's granularity representation in modeling coordinates may be non-uniform. The load imbalance is born here.

If, at time t_i, each processor possesses a fixed dataset, E_r, but looses a portion, E_l, from the clip operation and gains a portion of a neighbor's, E_g, at the same time, a deficit or surplus processor load balance arises, termed E_{ds} which is the net difference between E_g and E_l. The instantaneous per processor display set size $E_r(t_i)$ becomes:

$$E_r(t_i) = E_{ds}(t_i) + E_r(t_{i-1}); \text{ where } i \text{ denotes the simulation frame number, and } E_{ds}(t_i) = |\,E_g(t_i) - E_l(t_i)\,|. \quad 7.11$$

This equation states that the remaining processor load $E_r(t_i)$ at each instant of the simulation equals the previous resident quantity plus the absolute deficit/surplus amount resulting from display primitive set granularity and message-passing. This equation can be recast as:

$$E_r(t_i) - E_r(t_{i-1}) = E_{ds}(t_i); \text{ Or, } \frac{dE}{dt} = E_{ds}(t). \text{ The time rate of change for the per processor load balance is given by the deficit/surplus factor.} \quad 7.12$$

The global dead-time for each simulation frame T_f is proportional to the quantity MAX { $E_{ds}(t_i)$ } - MIN { $E_{ds}(t_i)$ }, for all processors. Each processor performs the identical rendering operations. The quantity of vertices to transform and render determines when each processor finishes. Locating the maximum or minimum deficit/surplus can be accomplished very efficiently in a hypercube via the bitonic sort algorithm invented by Batcher [Bat68]. The topology for the SCVS clearly favors a grid or mesh structure, and the overhead problem associated with implementing a logical hypercube topology within a mesh connection structure must be accounted for in the simulation. Alternatively, one could use a hypercube with node-resident hardware visualization support, but one must logically map each hypercube node into the tessellation.

The technique described here is termed *systolic compression-expansion* (SCE) load balancing. I have coined this term to account for the nature of real-time graphical rendering systems, where geometrical transformation often results in viewing the geometry from far distances (the compression mode), or up-close (the expansion mode). The display primitive set will tend to compress into a few tessellation elements (and the processor address spaces associated with each instance) when the distributed dataset is viewed from a long distance; compression mode forces the dataset to acquire a point-like appearance. When viewed up-close -- the expansion mode -- a few primitives will be replicated on many nodes. A terabyte-sized dataset can be visualized with the SCVS, but an attempt to view it from too far away will cause the interior-lying node address spaces to fill, and the outer nodes will have no work. A PRAM architecture (see Chapter 9) could be used to alleviate internal node filling in this case.

Concluding Remarks

Load balancing strategies for multicomputer simulations are varied and domain dependent. Simulation performance and multicomputer efficiency hinges on the load balance profile maintained during the course of execution, and is especially critical for dynamic and real-time domains. The simulation requirements dictate the load balancing strategy to use. The simulation's dynamic characteristics are derived from the data element mobility requirements. They impose frequency constraints on the load balancing operation.

The techniques presented in this chapter are representative of the load balancing methods introduced by investigators in response to the organization of multicomputer simulation methodologies appropriate to a particular problem domain. Many more load balancing strategies or variants of the methods discussed here will no doubt be introduced as new problems undergo multicomputer simulation. With respect to simulated annealing, the procedure can be executed in parallel on the multicomputer. A scheme for executing this *collisional* variant of the algorithm, and the consequences it holds for load balancing, is described by Williams [Wil90]. An analogous technique is described by Witte *et al.* [Wit91].

Suggested Reading

One of the earliest references describing the application of simulated annealing to a multicomputer simulation of a finite-element mesh problem is that of Flower, Otto, and Salama *et al.* [Flo86]. The reader is strongly encouraged to examine R.D. Williams [Wil90] for an elegant, and highly readable derivation of the simulated annealing cost-function for load balancing, and other algorithmic peculiarities.

Schwan [Sch88] has written a position paper on real-time parallel software applications which hints at the use of a hypercube for robotics control. One of the most fantastic and recent reports of a real-time multicomputer simulation applied to image processing can be found in Churbuck [Chu91] who reports its use in the next generation of cruise missiles. The paper by Jung *et al.* [Jun90] is a refined example of hybrid multicomputer architecture (both MIMD and SIMD processor elements are used) and real-time software for image processing. Crockett and Orloff [Cro91] have devised a rendering algorithm for MIMD-class architectures which uses an inspection method for data decomposition and load balancing.

The Southampton Novel Architecture Research Centre [Gle91] has published proceedings on recent directions in load balancing and performance monitoring for MIMD computers.

Raine *et al.* [Rai89] has developed a "systolic loop" load balancing mechanism appropriate for certain classes of molecular dynamics simulation.

The IOS Press (Amsterdam, Tokyo, and Washington D.C./Virginia) publish the Proceedings of the World Transputer User Group (WOTUG), Occam User Group (OUG), and the North American Transputer User Group (NATUG) which contain numerous examples of real-time multicomputer simulations for robotics, image processing, and other high-performance real-time problems.

Leslie Greengard has practically revolutionized the study of n-body problems for a specific class of conservative potential energy fields [Gre87].[11] His algorithm reduces the number of computations from $O(n^2)$, to $O(n)$ via a clever partitioning of the geometry, and the exploitation of the properties associated with partial wave expansions based on orthogonal polynomials.

---

[11] No extension of Greengard's algorithm to the Poisson-Boltzmann equation has been developed. This equation is used to generate the surface of potential energy for a molecule in solution. This is particularly useful for pharmaceutical modeling, with drug activity strongly influenced by the electrical interactions between molecules.

8

Synchronization

To achieve a predictable real-time simulation, the program's output must arrive in timely fashion according to granularity and strictness deadlines. Real-time simulations which satisfy these constraints embody predictability, a hallmark of successful real-time software engineering processes. Verification of a real-time simulation's predictability is conducted through the repeated observation of temporally and logically consistent processing. The concept of time is fundamental to predictable execution; all events and processing occur in relationship to a clock. In the multicomputer environment, N clocks, one for each processor, tick independently. The problem of establishing a multicomputer-wide time standard, against which all events are timed and registered to produce a temporally consistent distributed simulation, arises from this distributed architecture. This chapter discusses one approach to clock synchronization in distributed systems. Fault tolerant methods are purposefully excluded the discussion, though references to several algorithms are given (see Suggested Reading).

8.1 Background

A real-time simulation generates results which are temporally dependent. The time stamp assigned to each result is termed an *event*. Events are used to signify that a pertinent simulation state or condition has been achieved. They mark the passage of a simulation's progress. In sequential systems (including shared-memory multiprocessors), time stamps are derived and generated by a single clock, and events are logged according to simulation time, which is derived from the platform's clock source through periodic sampling of the clock output. In Chapter 5, we elaborated how the executive control structure fidelity depends on the clock granularity. A finer clock granularity supplies more accurate timekeeping; clock ticks arrive more frequently with increasing granularity.

Timekeeping via clock signals derived from a crystalline oscillator source is fundamentally imprecise since their capacity to generate ticks at a uniformly constant rate is limited. Their frequency output oscillation characteristics are known to drift in response to thermal perturbations and other physical properties of their structure. The stability of the clock signal derived from a crystal oscillator, $C(t)$, is often described by the following formula, which assumes that the clock signal is a continuous differentiable function:

$$\frac{dC(t)}{dt} \approx 1 \; ; \; or, \; \left| \frac{dC(t)}{dt} - 1 \right| < \epsilon \qquad \textbf{8.1}$$

where $\epsilon \approx 50 \times 10^{-6}$ is typical for crystal clocks.

The quantity ϵ represents the clock's drift.[12] At 1 MHz oscillation, $1.0 \pm 50 \times 10e^{-6}$ seconds elapse for a crystal clock; this drift amounts to a theoretical maximum drift ϵ_{max} of 1 millisecond over a 20 second interval. Drift is treated as a first-order effect, where changes in the clock's tick rate vary linearly with time; second-order effects are not considered.

A distributed simulation becomes synchronized to a master clock C_M at time t when, for all local clocks C_L on separate processors L, the following condition is satisfied:

$$| C_M(t) - C_L(t) | \le \gamma \text{ (seconds), for all processors } L. \qquad \textbf{8.2}$$

This inequality was first developed by Lamport [Lam78] who, in a seminal paper on the subject of event synchronization in distributed systems, established the theoretical basis for "modern" clock synchronization technique and method. The quantity γ is termed the *offset*, it is used to quantify the magnitude of the temporal agreement -- the synchronization -- between clocks *M* and *L*. The master clock process C_M generates time stamp data used to periodically adjust each node's local clock time reference. During the course of a distributed real-time simulation, the master clock process transmits to each node, at an interval of δt seconds, a reference time stamp value. The sequence of master clock updates to all local nodes occurs at times $t_k = k*\delta t$, where $k = 1, 2, ..., K$ intervals. The value of δt, the update interval, is fixed for each simulation. The *k*th resynchronization interval is defined as $[k*\delta t, (k+1)*\delta t]$. The local clock processes C_L on each processor *L* will have their references adjusted at the *(k+1)st* interval. Each local clock will indicate the time values $C_L(t_{k+1}) = C_M(t_k + \delta t)$, where $t_{k+1} = t_k + \delta t$ accounts for the elapsed simulation time, assumed to correspond to real-time accumulated since the simulation started ([Ram90], Copyright © 1990, The IEEE).

The coordination of a distributed simulation's temporal properties via this periodic maintenance process gives rise to the *time-stepped* simulation, a byproduct of *conservative* simulation methodology. All processes of the simulation are executed in parallel at time t_1, at t_2, and so on. Processes will run in this fashion until a deadlock condition arises. A monitoring process periodically checks for deadlock conditions, and will transmit a message to break the deadlock condition and resume the simulation.

The time-stepped simulation differs from a discrete-event simulation, where

---

[12] A value of $\epsilon = \pm$ 50 parts-per-million drift per each elapsed second of real-time is typical for non-thermally stabilized crystal clocks operating at 1 MHz. Atomic clocks are far more stable, with ϵ less than $\pm$ 0.01 parts-per-billion.

an irregular temporal aspect of the computational domain precludes a globally sequential time-step. In circuit simulation for instance, parts of an arithmetic logic unit (ALU) may require many more time-steps to complete an instruction cycle than another part of the cpu. When the circuit simulation is parallelized to speed up chip design evaluation, each portion of the circuit is distributed to a separate processor, and the timing properties of the simulation must be closely monitored and regulated during each simulated time period to locate propagation delays, pipeline stalls, and other asynchronous events.

Discrete event simulations often rely on an *optimistic* simulation methodology. An optimistic simulation executes asynchronous contexts and stores previous machine states in case the machine must rollback to a prior state. Processes execute and communicate until a time stamp is discovered that is earlier than the globally recognized simulation time. Some processes may compute far ahead of others, in simulated time, hence the term 'optimistic.' Rollback requires the cancellation of messages (via anti-messages) and a return to an earlier machine state that matches the anachronistic time stamp. The Time-Warp Operating System and the concept of virtual time (see Suggested Reading) embody this optimistic framework; it is a stark contrast to the conservative methodology most often practiced in real-time environments.

The MIMD-class time-stepped distributed simulation is analogous to a SIMD-class multicomputer computation, where a single instruction arbiter/controller simultaneously broadcasts program instructions to all processors in the platform. Each instruction in the SIMD broadcast causes simultaneous execution; the simulation progresses synchronously with the broadcast. In the MIMD model, each processor is autonomous and free to execute an instruction stream asynchronously from other peers. SPMD programs become SIMD congruent when each processor executes the identical instruction at the same time. The load balance of each processor, if perfectly matched for all time, necessarily implies this SIMD congruency.

The flexibility of MIMD multicomputers lies in their asynchronous capacity, which tolerates disparate processor loads, and forces synchronization among the asynchronous processes during message-passing. In the presence of dynamically varying processor loads, instruction level synchronization is lost and *weak* temporal coordination of the simulation emerges by default. Temporal coordination is therefore defined in terms of application-level synchronization. In contrast, a simulation with a frame time T_f nearly equal that of the processor's instruction cycle demonstrates strong temporal coordination. *Strong* temporal coordination dictates tightly coupled execution, and is supplied by true SIMD-class architectures. A real-time MIMD simulation with a frame time T_f assumed to be many milliseconds long demonstrates weak temporal coordination. The asynchronous characteristics of the simulation dominate over this period. During the interval T_f, process peers block and unblock at will since disparate load balance conditions prevent synchronous execution. So long as the dead-time T_d introduced from the load balance does not cause the quantity $T_f + T_d \pm \gamma$ to become larger than some bound D (see Chapter 7), the simulation will remain predictable.

The appearance of a globally accurate time reference at the local scale, with occasional maintenance, supplies weak temporal coordination. A real-time SPMD simulation which relies on weak temporal coordination must obey predictability constraints (see Chapter 5). Performance is still dictated by the most heavily laden node; events on other nodes occur in relationship to the slowest. The clocks C_L will advance independently and synchronously over short intervals while the simulation processes block until ready to proceed.

An event *X* recorded at simulation time *T* carries the uncertainty value γ, and is thus resolved as an event at time $T \pm \gamma$. As $\gamma \approx T_f$, the event record looses validity; the clock synchronization error is larger than the interval needed to complete one dispatcher cycle. Either γ must be reduced, or T_f increased to improve the event record resolution. The quality of the global temporal reference, the *skew* between any two clocks *i* and *j*, is inversely proportional to γ, and is stated by Dunigan [Dun91] as:

$$\phi_{ij} = \frac{\theta_{ij}(t + \delta t) - \theta_{ij}(t)}{C_i(t + \delta t) - C_i(t)} \qquad 8.3$$

where $\theta_{i,j}(t) = C_j(t) - C_i(t)$ is the offset between two clocks *i* and *j*. This equation relates the rate of change of offset with respect to a fixed interval of time; it quanitifies how fast the clocks drift apart during the interval.

The event record sequence comprises a simulation's history. The evolution of the event sequence is the primary record that chronicles the simulation. Load balance anomalies produce a *loosely time-synchronous* (what Fox terms 'loosely synchronous') distributed simulation. Loosely time-synchronous simulations are inherently asynchronous. The simulation proceeds unevenly, perhaps because of huge load balance oscillations. The load balance profile may exhibit chaotic behavior. The event record will show large and irregular temporal disparities between events rather than a regular or periodic pattern.

In the real-time SPMD environment, clocks tick independently though they are periodically synchronized, and reliance (or luck) is placed on the simulation's dynamic load balance profile to not violate predictability constraints. The purpose of clock synchronization in a real-time SPMD environment is to support event record logging that employs a consistent view of time across the multicomputer.

Messages are used to communicate the master clock reference in a distributed system. A local clock process C_L acquires the master reference signal, and adjusts its local reference in accordance with the transmission delay of the sender. Since messages are subjected to routing delays within communication networks, they do not arrive instantaneously when sent from C_M to C_L. Each hop[13] a message takes between the time stamp origin and the destination increases the message delay. Other forms of communication traffic introduce further delay, since the network may not be entirely free to conduct only temporal synchronization, but may also be required to carry messages pertaining to the simulation proper, such as the intermediate results generated by the simulation's process peers. The additional message traffic may also

---

[13] See section 6.3.2 on routing.

vary dynamically, in response to load conditions, and this fact will introduce greater clock skew.

During synchronization, the master clock reference value C_M received by the local clock process C_L at synchronization time t_k on node L is functionally described by Equation 8.4. This equation shows the relationship between the master clock reference value at time $C_M(t_k)$ and the elapsed real-time as recorded by node L upon arrival of the time stamp. The Carlini-Villano Synchronization Method discussed next makes use of the notion that message transmission is not instantaneous.

$C_L(t_k) = C_M(t_k) + \Delta = C_L(C_M(t_k) + \Delta)$, where Δ accounts for the message-passing delay between the master and the local clock. The clock time at the master will be measured as $C_M(t_k) = C_M(C_M(t_k) + \Delta)$ when C_L receives the timestamp packet. **8.4**

Adapted from [Ram90]. Copyright (c) 1990, The IEEE.

8.2 The Carlini-Villano Synchronization Method

One of the most straight-forward techniques for clock synchronization in distributed systems has been devised by Carlini and Villano (CV) [DCV88]. The CV method relies on the existence of a *Hamiltonian* path within the distributed network. A Hamiltonian path is a graph that connects each node exactly once. Figure 8.1 shows a with toroidial topology network, an example of a Hamiltonian path.

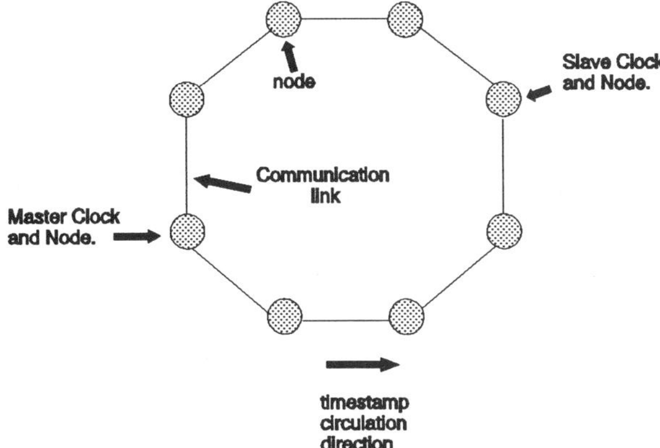

Figure 8.1 A toroidal multicomputer topology defines a Hamilton path.

Hamiltonian circuits can be found for hypercubes, grids, and other topologies. Synchronization messages that traverse a fixed multicomputer circuit will require a traversal time, T_{sync} (seconds), to orbit the circuit once. Message-passing times in fixed circuits can be reliably and repeatedly measured, and this is the principle reason for using them. Packet switching networks that employ wormhole routing, such as those formed by the SGS-Thomson/Inmos C104 routing chip, can be configured to accommodate a fixed path for specific virtual channels between process peers. Though it may be more expedient to perform L-1 wormhole routing messages (for L-1 slave clocks) as direct point-to-point messages from the master node M and clock C_M, provided that the message traversal time from M to L and back is bounded.

The CV method performs clock synchronization in multicomputers that are fault-free; their method is not fault tolerant, and this ascribes a simplicity and purity not found in more rigorous and costly techniques. If a real-time multicomputer clock fails, the results generated by a faulty clock immediately become suspect, and will likely jeopardize the entirely simulation.

According to CV, clock synchronization occurs at discrete intervals, denoted by Equation 8.5:

$$t_i = t_0 + i * \delta t, \textit{ for } i = 0, 1, 2, \ldots \textit{ and } t \geq t_0. \qquad \textbf{8.5}$$

where t_0 is the simulation's initialization time, and δt is defined as the time (in seconds) between synchronization messages. The master clock process C_M resides on processor M, and at intervals of δt seconds, broadcasts a time stamp packet $C_M(t_i)$ to all nodes L in the Hamiltonian path. The time of the broadcast t_i, and the receipt of the packet by the process that controls the clock C_L will occur at time $t_{arrival} = t_i + \Delta_{i,L}$, where $\Delta_{i,L}$ is the interval between the time of the broadcast and receipt of the time stamp packet (the message-passing delay) by the Lth clock process on node L during the ith synchronization interval.

The accumulated error during the synchronization interval bounded by $[\,t_i + \Delta_{i,L}, t_{i+1} + \Delta_{i+1,L}\,]$ is a composite of two quantities, one is a constant that results from the message-passing delay, and the other arises from clock drift and other perturbations. The synchronization error, $\gamma(t) = \gamma_L + \gamma_d$, where γ_L is the delay error, and γ_d is the drift error. The synchronization error at time t_i is defined as $\gamma(t_i) = \Delta_{i,L} + \Delta_{i,d}$, where $\Delta_{i,d}$ is the error caused by clock drift. The clock drift $\Delta_{i,d}$ is assumed to be identical for all clocks L, which implies that all clocks are the same make, model, and configuration. The CV method further requires that the drift term $\Delta_{i,d}$ at beginning of the interval is zero, i.e. that $\gamma(t) = 0$, for $t = t_i + \Delta_{i,L}$ which implies that the master and slave clocks are identically synchronized at the beginning of each resynchronization interval and merely drift apart between intervals. The CV algorithm corrects for both the delay and drift terms.

The delay term can be easily accounted for and removed from the synchronization error, since it is constant. At simulation time t_0, when initialization takes place, one can estimate the message-passing delay between the master M and the slave clock L by the quantity $\Delta^*_{0,L}$. The delay error at time t_0 will be given as $\gamma_L(t_0) = |\,\Delta^*_{0,L} - \Delta_{0,L}\,|$. Thus, one can derive the message-passing delay between the master node M and a slave L, by marking the time stamp prior to transmission slave L, and then waiting for the packet to return to the master. The time recorded by the master node M upon receipt of the packet from L will differ by the message-passing delay time, assuming the time needed to adjust the clock on node L is small compared to the transmission time. This *ping* activity may be repeated a few times to obtain a stable value for the message-passing delay Δ_L during initialization.

If a stable value of Δ_L is known, then the time to reach the Lth node in a Hamiltonian path will be given by $L*\Delta_L$, since the node sequence [0, 1,..., L-1, L, L+1, ..., N-1] are traversed in order, and the message delay for the time stamp to pass between successors is Δ_L. Carlini and Villano term Δ_L the partial delay of a 'single hop' -- the time required to send the time stamp packet between successor nodes in the Hamiltonian path. The estimate for initial delay required to forward a packet from the master M to slave L is denoted as $\Delta^*_{0,L}$, and can thus be assigned the value of $L*\Delta_L$. A more accurate value for Δ_L among N processors in the Hamiltonian circuit can be from $\Delta_L = t_{ring}/N$, with t_{ring} the time required for the time stamp synchronization packet to complete cycle of the Hamiltonian circuit during the synchronization interval.

During successive synchronization intervals $[t_i, t_{i+1}]$, the message-passing delay time Δ_L for the next synchronization interval $i+1$ can be refined as $\Delta_{i+2,L} = (t'_{ring} - t_{ring})/N$. Here, t'_{ring} is the time consumed by the time stamp packet to orbit the circuit at synchronization period t_{i+1}. If the message-passing network is not unduly loaded, $t'_{ring} = t_{ring} + N*\Delta_{i,L}$. The orbit time t_{ring} is influenced by the dynamic load balance, which is assumed to slowly varying in comparison to t_{ring}. A synchronization algorithm based on the CV method should organize and limit the synchronization period δt to a magnitude that is many times greater than T_f. If $\delta t \approx T_f$, with $T_f \approx$ 30 ms, a large amount of network bandwidth will be consumed by the synchronization activity, an invasive and superfluous protocol.

The synchronization period δt can be broadened if the drift factor Δ_d is minimized between synchronization intervals $[t_i, t_{i+1}]$. The clocks drift apart between synchronization periods. By using a predictive or extrapolation method, the slave clocks can keep amazingly accurate time between synchronization intervals. Carlini and Villano show results for extrapolation techniques applied to clock drift estimation under various loading conditions. Their experiment shows that a simple first-order linear extrapolation technique is generally superior (less arithmetic with nearly the equivalent accuracy) to quadratic prediction. An Occam version of the extrapolation algorithm and the CV "Ring_Sync" synchronization algorithm is given in [Car88].

Concluding Remarks

Faulty clocks cannot maintain a time standard within the bounds established by Equation 8.2, and algorithms for synchronizing in their presence are discussed by numerous authors, including Tannenbaum ([Tan92]) who describes the Lamport synchronization algorithm [Lam78], and Gusella [Gus89]. When a multicomputer node suffers a clock failure, all real-time processing will cease to become valid, and the load assigned to that processor must be redistributed. Fault tolerant multicomputer systems, like the CM-5 from Thinking Machines, Inc., the KSR1 from Kendall Square

Research, or the Parsytec GC permit nodes to be arbitrarily removed from the computation system when a failure is detected. The maintenance of real-time fidelity in the presence of failures effecting the most fundamental processor components is the true mark of successful engineering practice and discipline.

The CV technique is simple to implement and requires a minimum of messages to operate. It is these two elements which make their technique a strong candidate for real-time clock synchronization. One need only allow the clock processes on each node to run periodically at a higher interrupt priority level than the simulation processes to ensure low latency among the clock synchronization processes.

Suggested Reading

In contrast to the conservative simulation methodology, the optimistic methodology developed by Jefferson and Swizoral called the Time-Warp Operating System ([Jef85], [Jef88], [Jef90], [Fuj92]), and since extensively studied for distributed applications, provides a startlingly brilliant and elegant alternative for multicomputer systems. Some of the earliest comparisons and discussions of TWOS are presented in [Ung88]. The Chandy-Misra approach ([Cha81]) is the benchmark foundation for conservative simulation practice. Lipton and Mizell ([Lip90]) present a comparison of the conservative and optimistic simulation methodologies which proves that the Time-Warp mechanism can out-perform Chandy-Misra under certain conditions.

The sliding-window fault-tolerant clock synchronization algorithm by Pfluegl [Pfu91] utilizes a probabilistic synchronization technique, which is applicable to Byzantine faults with random message delays and random clock drifts. A patent was issued to Halpern *et al.* [Hal86] on a decentralized clock synchronization method. Ramanathan etal. provides an excellent overview and classification of fault-tolerant clock synchronization including both hardware and software techniques. Cole and Foxcroft ([Col88]) report on a clock synchronization experiment between computers located in the US and the U.K. French [Fre89] outlines a clock synchronization mechanism for hypercube multicomputers based on spanning trees.

9

Advanced Topics

This chapter explores the notion of parallel random access machine (PRAM) architectures posited by some computer architects to become the next computational paradigm for multicomputer platforms. Current multicomputer architectures complicate the software engineering process: topology, data decomposition, load balancing, synchronization, and other factors synthesize a difficult environment for software development purposes. To solve difficult problems like the Grand Challenges, the costs associated with designing and debugging the simulation software must become more expedient, and free the software engineer from wrestling with most of these issues, just as the von Neumann architecture allows for sequential platforms. The invention of a general purpose parallel processor which can be simply programmed like a standard personal computer or workstation would truly revolutionize the parallel computing field. PRAM architecture, while very embryonic, offers the potential for software engineers to hurdle the concurrent complexity chasm (see Skillcorn, [Ski90]).[14]

9.1 The Success of von Neumann[15]

With few exceptions today, most engineers and scientists employ a computer of some variety to assist them with their research or work efforts. The fundamental architecture used by computers is based on the work of von Neumann (Goldstine, [Gol80]). The von Neumann architecture simplifies software engineering and program development at several levels. The memory unit, where program and data storage reside, are under the control of a separate control entity, the processor. The two are interconnected through a bus. The processor acquires program instructions through a memory address generation mechanism over the bus, and obtains the instructions, or opcodes, along with the data. The opcodes direct the processor to operate on the

[14] This article provides a lucid overview of PRAM technology, definitions, and findings for this paradigm.

[15] Historically, John Vincent Atanasoff is credited with building one of the first electronic digital computers with separate memory store and processor components [Mac88].

data within the memory address space, or within the processor's registers. This sequence of control, process, and input/output operations serves as the foundation for all commonly available computer system platforms based on microprocessors. The architecture allows a sequential and predictable stream of instructions to move from memory to the processor and back again to memory.

The von Neumann architecture physically embodies the universality concept ascribed to Turing machines: "It is already explicit in Turing's fundamental work (Turing, [Tur36]) that a computer may be viewed both as a special purpose device executing a particular program, as well as a universal device capable of simulating all programs. The importance of this duality in sequential computation is enhanced by the fact that universality can be made efficient. This means that special purpose machines have no major advantage since general purpose machines can perform the same functions almost as fast. Efficient universality also makes possible general purpose high-level languages and transportable software, both of which contribute decisively to the ease of use of computers (Valiant, [Val90a])."

Modern programming languages, compilers, linkers, and operating systems create a friendly environment for software development activities. This support software further isolates the underlying memory and processor hardware control structures from the user. The translation between logical, symbolic program addresses to the physical machine addresses through the compilation and linking process substantially eases programming activities. The compiler's translation mechanism automatically generates physical addresses for data and code from the software abstraction expressed by the programming language. The processor begins sequential opcode load and store operations, and the program carries out the simulation when the executable image, the final instruction stream, is installed into memory. The term *general purpose processor* (GP^2) is attributed to computer systems of the von Neumann architecture class, and arises, in part, from the ease of program address translation afforded by compilers and other tools designed to generate a sequential instruction stream.

The term "general purpose" implies that one can apply the architecture with equal effectiveness toward the solution of all problems expressible by a programming language. Variants of the architecture are employed to obtain faster solutions or for less cost. The GP^2 sequential address translation scheme is comfortable for human minds. Contemplating one action at a time mirrors the reality found in our daily existence. The instruction stream generated by a compiler for general purpose sequential processors, with some exceptions,[16] causes one operation at a time to occur within the processor.

The sequential paradigm is the basis for almost all commercially available computer systems. In Chapter 1, a heuristic derivation of the expected solution speedup showed that a sequentially structured shared-resource computer system (a multiprocessor) possesses an asymptotic performance characteristic. Shared-resource systems exhibit this limiting behavior; they are contention bound. Alternatively, the same heuristic derivation demonstrated how a non-shared resource computer system,

---

[16] Vector processing units and pipeline constructs are notable exceptions.

systems exhibit this limiting behavior; they are contention bound. Alternatively, the same heuristic derivation demonstrated how a non-shared resource computer system, a multicomputer, is capable of linear speedup. Multicomputers request information exchange via message-passing, and are contention-free; shared-resource systems are contention-bound by design.

Multicomputer architectures are also based on the von Neumann paradigm, where multitudes of processors and memories, each executing a decoupled sequential instruction stream, are harnessed to solve more complex problems. The instruction stream may invoke message-passing operations, which are analogous to a remote logical-to-physical address decode and lookup function. Message-passing serves as a surrogate supplement to the sequential instruction stream. The memory hierarchy decomposition imposes new challenges for the software engineer to account for the distributed address spaces. Instructions and data are no longer expediently located in a conveniently reachable global address space. This complication, a wrench in the gears of standard logical-to-physical address translation, has so far prevented the organization of an equivalent *general purpose parallel processor* (GP^3).

While based on multiple instances of GP^2 architecture, current GP^3 architectures do not provide the comfort and familiar problem solving paradigm attributed to von Neumann. Chapters 6 through 8 of this text demonstrate the architectural and software engineering disparities between the two paradigms. The address translation mechanism -- the software support -- serves as the foundation for a general purpose computer system. The hardware is only along for the ride. A GP^3 cannot exist, for all practical purposes, without a topologically independent address translation scheme.

Message-passing primitives, physical concurrency definitions for computational domain topology and routing, and explicit data decomposition operations represent substantial obstacles which simultaneously block, but also sustain the current generation of multicomputer systems, engineers, and software projects. In 1990, multicomputer vendors grossed approximately US $170M[17] in sales, and this figure is likely to grow as scalable computation becomes ever more attractive through hardware cost/performance benefits, even if software engineering costs offset these gains. PRAMs need not be built from exotic silicon, the old hardware provides a useable foundation for PRAM operation. However, software is needed to sew the existing hardware into a seamless GP^3. A literal explosion in technology, a societal revolution and concomitant upheaval, will arise if the simplicity of programming a parallel computer can be achieved with the ease attributed to standard sequential von Neumann architectures. Many businesses will become far more efficient through the increased displacement of workers, since their electronic replacements will be far more powerful and flexible at solving problems.

---

[17] Projected at US $250M in 1992 [Mar92b].

9.2 PRAM Overview[18]

PRAM architectures seek to simplify and reduce current multicomputer system programming chores. Articulation of the machine architecture, and enhancement of software support tools (e.g., compilers and operating systems) are necessary to realize a universal GP^3 architecture and software environment. A multicomputer software engineer currently must implement explicit message-passing operators and exercise manual memory management (storage allocation through data decomposition) operations during program development. These operations require substantial effort, are not always easy or painless to perform, and therefore raise engineering costs. A successful PRAM architecture will automatically conceal these two highly visible attributes, and restore efficiency lost through the emergence of currently burdensome multicomputer architectures. The principle objective of PRAM technology hinges on the innovation and deployment of an address translation mechanism which hides the underlying machine topology and architecture from the software application. This mechanism supplies a bridge (Valiant, [Val90b]) between the von Neumann and parallel computation paradigms which will ameliorate the symptomatic complexities found in current multicomputer software engineering practice.

The address translation schemes currently employed for multicomputer software rely on manual memory management and data decomposition schemes arising from the inclusion of message-passing operators into the application. The message-passing operators account for the multicomputer's distributed memory and interconnection topology. The memory management component arises from the manual assignment and allocation of data and code to each processor. Current methodologies, such as scattered decomposition and simulated annealing, or the Linda tuple-space mechanism, ease the programming chore associated with memory management. The Linda system conveniently conceals both message-passing and memory allocation, and affords machine architecture independence for the application. Tuple-space operations occur transparently, and may randomly disperse the data throughout the multicomputer. Information recovery from tuple-space is equally transparent, and occurs via the invocation of inverse Linda operators.

The convenience and architectural independence supplied by Linda restores the most comfortable notion of the von Neumann paradigm. Linda's operators are manifested as simple programming language extensions which are transparently executed on top of the multicomputer. Linda is a harbinger of PRAM technology. But the topological, message-passing, and memory management transparency which Linda affords is *an* answer, but by no means *the* answer or the most general solution for bridging the sequential and parallel domains.

Linda is not entirely ubiquitous. Imagine the problem of processing a gigantic dataset. With several gigabytes or even a terabyte in extent, a gigantic dataset cannot be efficiently loaded into a distributed memory platform if the dataset resides on one

---

18 The discussion is confined to PRAM emulation via message-passing multicomputer systems. No consideration of butterfly, multistage, or bus-connected platforms is given.

mass storage device.[19] For efficiency considerations, the dataset must first be decomposed, then distributed, and finally loaded into Linda. The initial dataset decomposition can be achieved through the imposition of a structure or hierarchical organization. Alternatively, if the dataset is first collected from a distributed acquisition system, so much the better -- it is pre-decomposed. In either event, the structure must be known *a priori* to preserve program address coherency in an explicit and dependent message-passing environment.

The PRAM message-passing interconnection network must permit each processor to freely access any of the connected memories under program control via a simple address generation request. This implies that message-passing operators must be emitted by the compiler as a consequence of code generation and data dependency analysis. In the PRAM environment, the application software will not contain the explicit declaration of message-passing primitives and operators commonly found with existing applications (see section 6.7). This programming approach represents a radical departure from current practices. Accordingly, the compiler must possess the capability to automatically generate the instruction stream for non-local address references. The running program will issue references to data, and the references are somehow fulfilled by a *reference concordance mechanism* (RCM); explicit knowledge concerning the distribution of data within the multicomputer's memory structure may not be maintained by the compiler.

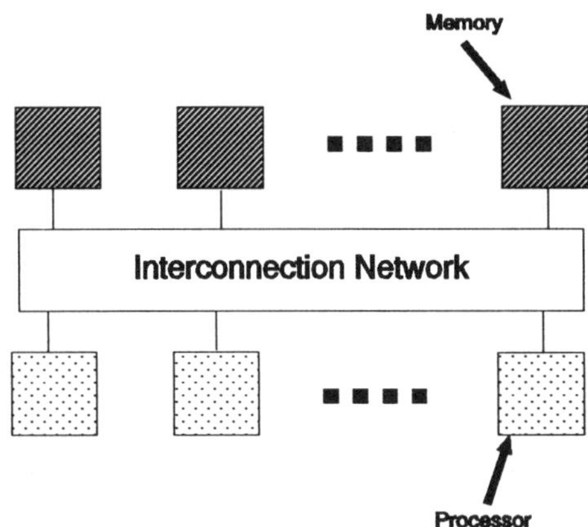

Figure 9.1 A PRAM architecture schematic.

---

[19] Standard mass storage systems are considered (e.g., hard-disks, tapes, etc.). The optical memory systems [Ste92b] being researched will change this contemporary view.

An RCM functions as a distributed server which translates the machine's topology, network structure, and data distribution into a coherent address translation operation through a functional abstraction. The RCM performs address resolution for reference requests routed within the PRAM interconnection network (see Figure 9.1). When a processor generates a non-local memory reference, the RCM translates the request into a packet which is then routed to the destination address space where the request is satisfied, and the data is automatically returned to the requesting processor for further operations. The RCM is analogous to Linda in this respect. When a non-local reference is issued, the packet destination must be known or decoded by the RCM, and this implies knowledge of the multicomputer topology. Alternatively, a broadcast operation can be used to locate the data, but at high cost in terms of network bandwidth consumption and accumulated processor time needed to decide if the address is local to a processor's memory. A multicast operation would be more efficient in this respect, since a few destination addresses are used for this function. The RCM must therefore have *a priori* knowledge of the dataset distribution, or an alternative efficient lookup capability for rapidly determining where to direct an address reference request.

9.3 Performance Issues

What software technology is necessary to imbue PRAM characteristics, and obtain a GP^3 architecture from an existing multicomputer platform? A mechanism which can serve as the address translation scheme must be engineered and incorporated into existing programming languages and operating systems. A technique known as *parallel hashing* has been developed for PRAM emulation which affords a simple mechanism for address translation for multicomputers. Parallel hashing relies on the properties of universal hash functions described by Carter and Wegman [Car79]. A universal hash function (UHF) supplies random addressing within a distributed memory hierarchy. The UHF possesses properties which are identical to the simple hash function in a single address space: minimum collisions and efficient computation. The random address generation mechanism supplied via universal hashing helps to eliminate access conflicts -- memory collisions -- between processors.

Karlin and Upfal [Kar88] have investigated the behavior of an Exclusive-Read-Exclusive-Write-PRAM (EREW-PRAM) simulation. The EREW-PRAM assumes that no program variable is simultaneously accessed by two processors. To simulate the PRAM architecture, they use a universal hash function defined as:

$$\begin{aligned} H = \{h \mid h(x) &= ((\sum_{0 \le i < \zeta} a_i x^i) \bmod p) \bmod n, \\ &\textit{for any } a_i \in [0 \cdots p-1],\ a_0 \in [1 \cdots p-1]\} \end{aligned} \qquad \textbf{9.1}$$

where p is a prime number, m the number of logical addresses, n the number of processors, and ζ = c $\log_2 m$ (c is a constant). For efficiency considerations, $m \gg n$. This hash function resides on all processors, and will likely become an embedded component of a PRAM multicomputer operating system. When a processor attempts to read or write a variable α, it computes $h(\alpha)$ to determine which processor it must communicate with and obtain the reference. The value computed from the UHF contains not only the processor identity, but the address within the non-local memory which must be read to obtain the requested information. Karlin and Upfal have shown that this hash operation consumes an additional O(log m) steps in the simulation of each PRAM instruction. If existing hardware implementations are used to instance PRAM architectures, a modest slow-down in the processing bandwidth occurs.

The critical issue associated with PRAM operation is the assignment and allocation of variables to processors and their address spaces in a random fashion through application of the UHF. If a dataset D is organized as a matrix of values, and we wish to employ the PRAM architecture for a computation, each value -- the logical address of D_i -- must be distributed according to the application of $h(D_i)$. The UHF generates a random processor index for the physical address assigned to the dataset value by the hash function. D_i is placed in the corresponding processor's address space.

For the maintenance of a deterministic simulation, each address reference request handled by the RCM must be synchronous and blocking. Since an O(log m) steps are consumed for each address reference, each processor will become idle for that period; dead-time will accumulate. To counter the intrinsic dead-time accumulation penalty, each processor should simultaneously execute O(log m) separate contexts. A PRAM architecture, like the existing multicomputer platforms, will best be utilized by overlapping communication with additional computation. The addition of multiple contexts to each PRAM processor conceals the dead-time through the concurrent issue of address reference requests and overlapped computation.

Concluding Remarks

Given the revolutionary advances occurring in the computing field, it is fairly safe to say that PRAM technology, at some level of functionality and capability, will probably arrive on the desktop within 5 years. The standardization of microprocessors, programming languages, and operating systems will further the rapid rise of PRAM technology and other off-shoots of this branch.

It is very difficult for the author to conceive and visualize how PRAM technology may influence the SCVS architecture. On the one hand, the data decomposition and message-passing operations are entirely eliminated from the software engineering process, and this benefit is highly attractive, especially for potential users of the SCVS who are not multicomputer software engineering experts. Alternatively, acceptance of this transparent vehicle complicates the conceptual understanding of frustum decomposition and communications aspects associated with real-time multicomputer simulation predictability criteria. Universal hashing would randomize a regular structure of data suitable for simple decomposition and visualization via the SCVS, and it is unclear whether or not this is beneficial or detrimental to an SCVS/PRAM marriage.

The marketplace is beginning to sponsor commercial PRAM development (Markoff, [Mar92b]). The KSR1 from Kendall Square Research (Waltham MA) has

(Markoff, [Mar92b]). The KSR1 from Kendall Square Research (Waltham MA) has developed a hierarchical memory structure and microcoded firmware control structure [Fra91] which implements a shared-distributed memory paradigm [And92], perhaps the first commercial instance of PRAM technology. A performance comparison of the the Intel iPSC-860 Touchstone and the KSR1 can be found in Dunigan ([Dun92]).

Suggested Reading

The Monsoon Dataflow Project, a joint MIT and Motorola research project, stands as a recent example of an experimental non-von Neumann architecture ([Tra91], [Pap91], [Nik89]). Dataflow architectures afford a realistic and useful alternative computing paradigm. Natvig ([Nat89a] and [Nat89b]) has built a PRAM simulator based on the SIMULA programming language. The Parallaxis system devised by Braunl [Brau90] for SIMD architectures emulates a PRAM architecture, but relies on the embedding of topological specifications and descriptions within the program body. Mehlhorn and Vishkin [Meh84] have proved some very important results for PRAM address translation mechanisms and simulation. Although not discussed here, systolic computation, as embodied by the iWARP system developed at Carnegie Mellon University, and since commercialized by Intel is distinguished as a special purpose computing architecture with carefully designed and powerful software support tools ([Hal91], [Bok90], [Sub91], [Web91]).

Bibliography

[Ahu86] S. Ahuja, N. Carriero, and D. Gelertner, "Linda and Friends," *IEEE Computer*, Vol. 19, No. 8, August, 1988, pp. 26-34.

[Alm89] George S. Almasi and Allan J. Gottlieb, Highly Parallel Computing, Benjamin/Cummings Publishing Company, Inc., Redwood City, CA 1989.

[Amd67] G.M. Amdahl, "Validity of the single processor approach to achieving large-scale computing capabilities," *AFIPS Conference Proceedings 1967*, 30, AFIPS Press, Montvale NJ, 1967, pg. 483.

[And91] E.L. Andrews, "String of Phone Failures Perplexes Companies and U.S. Investigators," *The New York Times*, July 3, 1991, pg. A1. "Computer Maker Says Flaw in Software Caused Phone Disruptions," *The New York Times*, July 10, 1991, pg. A10. "The Precarious Growth of the Software Empire," *The New York Times*, July 3, 1991, Section 4, pg. 1. See also "Three little bits breed a big, bad bug," by G.F. Watson, *IEEE Spectrum*, pg. 52, May 1992.

[And92] E.L. Andrews, "Simplified Software for Computer Systems," *The New York Times*, Nov. 30, 1991, pg. 36. Discusses U.S. Patent # 5,055,999 by H. Burkhardt of Kendall Square Research. Patents available for US $3 from the Patent and Trademark Office, Washington D.C. 20231.

[Arf70] George Arfken, Mathematical Methods for Physicists, 2nd Edition, Academic Press, New York, 1970.

[Ath88] W.C. Athas and C.L. Seitz, "Multicomputers: Message-passing concurrent computers," *IEEE Computer*, Vol. 21, No. 8, pp. 9-24, August, 1988.

[Aue86] Brent Auernheimer and Richard A. Kemmerer, "RT-ASLAN: A Specification Language for Real-time Systems," *IEEE Transactions on Software Engineering*, Vol. 12, No. 9, Sept. 1986, pp. 879-899.

[Bar82] E.R. Barnes, "An Algorithm for Partitioning the Nodes of a Graph," *SIAM J. Alg. Disc. Meth.*, Vol. 3, pp. 541, 1982.

[Bar91] J. Barron, "Catching Times Sq.'s Eye, in a Flash," *The New York Times*, Section 1, Jan. 16, 1991.

[Bat68] K.E. Batcher, "Sorting networks and their applications," *AFIPS Conference Proceedings 1968*, 32, pg. 307, AFIPS Press, Montvale, NJ 1968.

[Bea90] T. Bearden, "Focus -- Software Safety," *MacNeil/Lehrer News Hour*, February 21, 1990, Show #3673. Transcripts available for US $4 from Strictly Business, P.O. Box 12361, Overland Park, Kansas, 66212.

[Ber66] A.J. Bernstein, "Analysis of Programs for Parallel Processing," *IEEE Transactions on Electronic Computers*, Vol EC-15, No. 5, Oct. 1966.

[Blo89] J.J. Bloombecker, "Malpractice in IS," *Datamation*, pp. 85-86, Oct. 15, 1989.

[Boe81] B.W. Boehm, Software Engineering Economics, Prentice-Hall, Englewood-Cliffs New Jersey, 1981.

[Boe86] B.W. Boehm, "A Spiral Model of Software Development and Enhancement," *ACM SIGSOFT Software Engineering Notes*, Vol. 11, No. 4, Aug. 1986.

[Bok90] S. Bokhari, "Communication Overhead on the Intel iPSC-860 Hypercube," Institute for Computer Applications in Science and Engineering, NASA ICASE Report 10, NASA Langley Research Center, Hampton, VA, 23665-5225.

[Bop87] R.B. Boppana, Eigenvalues and Graph Bisection: an Average Case Analysis, in 28th Annual Symp. Found. Comp. Sci., 1987.

[Bor90] Shekhar Borkar, et.al., "Supporting Systolic and Memory Communication in iWARP," *Proceedings of the 17th Annual International Symposium on Computer Architecture*, Seattle WA, May 28-31, 1990.

[Bou77] William Bourke, Bryant McAvaney, Kamal Puri, and Robert Thurling, "Global Modeling of Atmospheric Flow by Spectral Methods," appearing in General Circulation Models of the Atmosphere, Julius Chang Ed., *Methods in Computational Physics*, vol. 17, 1977; Academic Press, New York.

[Bow92] Jonathan Bowen and Victoria Stavridou, "Safety-Critical Systems, "Formal Methods and Standards," Programming Research Group Technical Report PRG-TR-5-92, Oxford University Computing Laboratory, 11 Keble Road, Oxford OX1 3QD, England.

[Bow87] Ken C. Bowler, Alastair D. Bruce, Richard D. Kenway, G. Stuart Pawley, and David J. Wallace, "Exploiting highly concurrent computers for physics," *Physics Today*, Oct. 1987, pp. 40-48.

[Bra78] R. N. Bracewell, The Fourier Transform and its Applications, Second Edition, McGraw-Hill Book Company, New York, 1978.

[Brad89] Raymond Bradbury, Zen in the Art of Writing: Essays on Writing and Creativity, Joshua Odell Editions, Santa Barbara, CA, 1989.

[Brau90] Thomas Braunl, "The User Manual for Parallaxis Version 2.0," University of Stuttgart IPVR, Breitwiesenstrasse 20-22, D-7000 Stuttgart 80, Germany. Available for US $15 or DM 10. Also available via anonymous ftp from ftp.informatik.uni-stuttgart.de (129.69.211.1) in subdirectory pub/parallaxis.

[Brau91] Thomas Braunl, "The Open Channel -- Braunl's Law," *IEEE Computer*, August 1991, pg. 120.

[Brd91] W. J. Broad, "Space Errors Share Pattern: Skipped Tests," *The New York Times*, Section B, pg. 5, June 11, 1991.

[Bro75] F.P. Brooks, Jr., The Mythical Man-Month: Essays on Software Engineering, Addison-Wesley Publishing Co., Reading, Massachusetts, 1975.

[Bru90] F. Brugé and S.L. Fornili, "A distributed dynamic load balancer and its implementation on multi-transputer systems for molecular dynamics simulation," *Computer Physics Communications*, Vol. 60, pp. 39-45, 1990. "Concurrent molecular dynamics simulation of spinodal phase transition on transputer arrays," *Computer Physics Communications*, Vol. 60, pp. 31-38, 1990.

[Bur58] Robert Burns, *To a Mouse*, Stanza 1, 1758.

[Bur89] G.D. Burns, "The Local Area Multicomputer," *The Third Conference on Hypercube Concurrent Computers*, (Now called the Distributed Memory Computing Conference, Monterey CA, 1989. Golden Gate Publishers, Los Altos California. (Out of Print).

[Bur81] R.L. Burden, J.D. Faires, A.C. Reynolds, Numerical Analysis, pg. 517, Prindle, Webber & Schmidt, Boston MA, 1981.

[Car91] Luis-Felipe Cabrera and Darrell D.E. Long, "Swift: Using Distributed Disk Striping to Provide High I/O Data Rates," UCSC-CRL-91-46, Dec. 1991, University of California, Santa Cruz CA 95064.

[Car79] L. Carter and M. Wegman, "Universal classes of hash functions," *J. Comput. Syst. Sci.*, Vol. 18, No. 2, pp. 143-154, 1979.

[Car88] N. Carriero, and D. Gelertner, How to Write Parallel Programs, A Guide to the Perplexed, Yale University Technical Report DCS/RR-628, May 1988.

[Cha81] K.M. Chandy and J. Misra, "Asynchronous Distributed Simulation via a Sequence of Parallel Computations," CACM, Vol. 24, No. 11, pp. 198-206, April, 1981.

[Cha88] Stephen S. Cha, Nancy G. Leveson, and Timothy J. Shimeall, "Safety Verification in Murphy Using Fault Tree Analysis," *Proc. of the International Conference on Software Engineering*, Singapore, 1988. IEEE (1988), pp. 377-386.

[Cha90] Long-chu Chang and Brian T. Smith, "Classification and evaluation of parallel programming tools," Technical Report CS90-22, Department of Computer Science, The University of New Mexico, Albuquerque, NM 97131.

[Che89] Peter P. Chen, "The Entity-Relationship Approach," *Byte*, April, 1989, pp. 230-232.

[Chu90] Yan Chunning and Shi Dinghua, "Classification of Fault Trees and Algorithms of Fault Tree Analysis," *Microelectronic Reliability*, Vol. 30, No. 5, pp. 891-895, 1990.

[Chu91] David Churbuck, "The computer as detective," *Forbes*, Dec. 23, 1991, pp. 150-155.

[Col88] R. Cole and Clare Foxcroft, "An Experiment in Clock Synchronization," *The Computer Journal*, Vol. 31, No. 6, 1988, pp-496-502.

[Coo83] James E. Coolahan and Nicholas Roussopoulos, "Timing Requirements for Time-Driven Systems Using Augmented Petri Nets," *IEEE Transactions on Software Engineering*, Vol. 9, No. 5, Sept. 1983, pp. 603-616.

[Cop90] David Cooper, "Educating Management in Z," pp. 192, Z User Workshop, Proceedings of the Fourth Annual Z User Meeting (Oxford, England), Springer-Verlag, London, 1990. ISBN 0-387-19627-7.

[Cor91a] E. Corchoran, "Flat Horizons," *Scientific American*, June '91, pp. 112-14.

[Cor91b] E. Corchoran, "Calculating Reality," *Scientific American*, Vol. 264, No. 1, pp. 101-09, Jan. '91. See pg. 108.

[Cro91] T.W. Crockett, and Tobias Orloff, "A Parallel Rendering Algorithm for MIMD Architectures," NASA ICASE Report 91-3, June 1991, Institute for Cmputer Applications in Science and Engineering, NASA Langley Research Center, Hampton, VA, 23665-5225.

[Cus91] M. A. Cusumano, Japan's Software Factories, A Challenge to U.S. Management, Oxford University Press, New York, 1991.

[Cush91] John H. Cushman, Jr., "Air Control Alarm is Called Flawed," *The New York Times*, Oct. 9, 1991, Section A, pg. 7.

[Dav91] Bob Davis, "Pentagon Unit Steers Supercomputer Deals To Certain Companies," *The Wall Street Journal*, August 6, 1991, pg. A1.

[DCV88] Ugo De Carlini and Umberto Villano, "A Simple Algorithm for Clock Synchronization in Transputer Networks," *Software--Practice and Experience*, Vol. 18, No. 4, pp. 331-347, April 1988.

[Del88] Norman Delisle and David Garlan, "A formal Specification of an Oscilloscope," Technical Report, CR-88-13, Computer Research Laboratory, Tektronix Inc., P.O. Box 500, Beaverton, OR 97077 USA, Oct. 1988.

[Dew84] A.K. Dewdney, "Computer Recreations," *Scientific American*, Dec. 1984.

[DMC90] See Session 40, Vol. I on **Performance Monitoring and Profiling** in *The Proceedings of The Fifth Distributed Memory Computing Conference*, April 8-12, 1990, Charleston, South Carolina. Available from the IEEE Computer Society Press.

[DMC91] See the **Performance Analysis Track** in *The Proceedings of the Sixth Distributed Memory Computing Conference*, April 28-May 1, 1991, Portland Oregon. Available from the IEEE Computer Society Press.

[DoD85] DoD Standard 2167, Defense System Software Development, United States Department of Defense Washington, D.C. 20301, 1985.

[DoD84] DoD Standard 882B, System Safety Program Requirements, United States Department of Defense, Washington, DC 20301, 1984.

[DoD83] DoD Standard 1679, Software Development, United States Department of Defense, Washington, DC 20301, 1983.

[Duk91] Roger Duke, Paul King, Gordon Rose, and Graeme Smith, "The Object-Z Specification Language (Version 1)," Technical Report 91-1, Software Verification Research Centre, The University of Queensland, Queensland 4072, Austrailia. Available from the Formal Methods Group library: fmg@cs.uq.oz.au through Internet.

[Dun91] Thomas H. Dunigan, "Hypercube Clock Synchronization," ORNL/TM-11744, Mathematical Sciences Section, Oak Ridge National Laboratory, P.O. Box 2008, Bldg. 6012, Oak Ridge, TN 37831-6367.

[Dun92] Thomas H. Dunigan, "Kendall Square Multiprocessor: Early Experiences and Performance," ORNL/TM-12065, Mathematical Scieneces Section, Oak Ridge National Laboratory, P.O. Box 2008, Bldg. 6012, Oak Ridge, TN 37831-6367.

[Fey88] R.P. Feynman, "What Do You Care What Other People Think?", Further Adventures of a Curious Character, pp. 232-236, W.W. Norton & Company, New York, 1988.

[Fla89] H. P. Flatt and Ken Kennedy, "Performance of Parallel Processors," *Parallel Computing*, Vol. 12, 1989, pp. 1-20.

[FHP89] The Federal High Performance Computing Program, Office of Science and Technology Policy, Sept. 8, 1989. Document # PB50-159823 available from the U.S. Department of Commerce, National Technical Information Service (NTIS), Springfield, VA 22161.

[Flo87] J.W. Flower, S.W. Otto, and M.C. Salama, "A Preprocessor for Irregular Finite Element Problems," *Proceedings, Symposium on Parallel Computations and Their Impact on Mechanics*, ASME Winter Meeting, 14-16 Dec., Boston, MA, 1987. Also, California Institute of Technology Concurrent Computation Program Note C^3P 292.

[Fox84] Geoffrey C. Fox and Steve W. Otto, "Algorithms for Concurrent Processors," *Physics Today*, May 1984, pp. 50-59.

[Fox88a] Geoffrey C. Fox, Mark A. Johnson, Gregory A. Lyzenga, Steve W. Otto, John K. Salmon, David W. Walker, Solving Problems on Concurrent Processors, Volume I, General Techniques and Regular Problems, Prentice-Hall, Englewood Cliffs, NJ, 1988.

[Fox88b] G.C. Fox in Numerical Algorithms for Modern Parallel Computers, ed. M. Schultz, Springer-Verlag, Berlin, 1988.

[Fra91] S.J. Frank, H. Burkhardt III, L.O. Lee, N. Goodman, B.I. Margulies, F.D. Weber, US Patent 5,055,999 issued Oct. 8, 1991 available from the Patent and Trademark Office, Washington D.C. 20031.

[Fre87] Peter Freeman, Tutorial: Software Reusability, IEEE Computer Society Press, Catalog #EH0256-8, P.O. Box 80452, Worldway Postal Center, Los Angeles, CA 90080 USA. See "The Draco Approach to Constructing Software from Reusable Components," by J.M. Neighbors on pp. 179-180.

[Fre89] J.C. French, "A global time reference for hypercube multicomputers." In Hypercube Concurrent Computers and Applications 1989, pp. 217-220, J.L. Gustafson Ed., Golden Gate Publications, Los Altos CA, 1990. This publication is not widely available, since the conference title is now called the distributed memory computing conference; the publisher seems to have disappeared. Neither the University of California or the California State University Library systems have this publication in their stacks.

[Fuj92] R. Fujimoto, "Design and Evaluation of the Rollback Chip: Special Purpose Hardware for Time Warp," *IEEE Transactions on Computers*, Jan. 92.

[Fur86] G.C. Fox, and W. Furmanski, "Optimal Communication Algorithms on the Hypercube," California Institute of Technology Concurrent Computation Program, C^3P 314, July, 1986.

[Gan89] Chris Gane, "The Gane/Sarson Approach," *Byte*, April, 1989, pp. 224-227.

[Ger76] S.L. Gerhart and L. Yelowitz, "Observations on the Fallibility in Applications of Modern Programming Methodologies," *IEEE Transactions on Software Engineering*, Vol. SE-2, pp. 195-207, 1976.

[Gle91] Load Balancing and Performance Monitoring for MIMD Computers, Ian Glendinning (ed.), SNARC 91-01, Southampton Novel Architecture Research Center, University of Southampton, Department of Electronics and Computer Science, Southampton, England, Sept. 11, 1991.

[Gol80] Herman H. Goldstine, The Computer from Pascal to Von Neumann, Princeton University Press, Princeton NJ, 1980.

[Grape] These references are attributed to [Kah91], and contain no author identities. "A special-purpose computer for gravitational many-body problems," *Nature Japan*, Vol. 345 No. 6270, pp. 33-35, 3 May 1990; "GRAPE: Special Purpose Computer for Simulation of Many-Body Problems," International Symposium on Supercomputing, Fukuoka Japan, Nov. 6-8, 1991; "Treecode with a Special-Purpose Processor, to appear, *Publications of the Astronomical Society of Japan (PASJ)*, 1991; "Project GRAPE: Special Purpose Computers for many-body problems," to appear in High Performance Computing: Research and Practice in Japan, R. Mendez, Ed., John Wiley & Sons, 1992; "A special-purpose computer for gravitational many-body systems: GRAPE-2," *PASJ*, Vol. 43, pp. 547-555, 1991; "GRAPE-2A: A Special Purpose Computer for Simulation of Many-Body Systems with Arbitrary Central Force," HICSS-25, 25th Hawaii International Conference on System Sciences, Koloa HI, Jan. 7-11, 1992; "GRAPE-3: Highly Parallelized Special-purpose Computer for Gravitational Many-body Simulations," *ibidem*. "A modified Aarseth code for GRAPE and vector processors," in press, *PASJ*, 1991; "A special purpose N-body machine GRAPE-1," *Computer Physics Communications*, Vol. 60, pp. 187-194, 1990.

[Gre87] Leslie F. Greengard, The Rapid Evaluation of Potential Fields in Particle Systems, ACM Distinguished Dissertation Series, The MIT Press, Cambridge MA, 1987.

[Gua88] S.B. Guarro, "A Logic Flowgraph-Based Concept for Decision Support and Management of Nuclear Plant Operation," *Reliability Engineering and System Safety*, Vol. 22, 1988, pps. 313-332.

[Gua90] S.B. Guarro, J.S. Wu, M. Yau, and G.E. Apostolakis, "Findings of a Workshop on Embedded System Software Reliability and Safety," UCLA-ENG 90-25, Mechanical, Aerospace and Nuclear Engineering Department, University of California, Los Angeles, CA 90024-1597 USA, June 1990.

[Gua91] S.B. Guarro, J.S. Wu, G.E. Apostolakis, and M. Yau, "Embedded System Reliability and Safety Analysis in the UCLA ESSAE Project," appearing in *Probabilistic Safety Assessment and Management*, G. Apostolakis, Ed., Elsevier Science Publishing Co. Inc., New York, 1991.

[Gus86] J.L Gustafson, G.R. Montry, R.E. Benner, "Development of Parallel Methods For a 1024-Processor Hypercube," *SIAM J. on Scientific and Statistical Computing*, Vol. 9 (4), July '88, pp. 609-41.

[Gus89] Riccardo Gusella and Stefano Zatti, "The Accuracy of the Clock Synchronization Achieved by TEMPO in Berkeley UNIX 4.3BSD," *IEEE Transactions on Software Engineering*, Vol. 15, No. 7, pp. 847-853, July 1989.

[Haj88] B. Hajek, "Cooling Schedules for Optimal Annealing," *Math. Oper. Res.*, Vol. 13, pp. 311., 1988.

[Hal86] J.Y. Halpern, B.B. Simons, H.R. Strong, "Decentralized Synchronization of Clocks," *U.S. Patent Number* 4,584,643, Apr. 22, 1986.

[Hal91] David R. O'Hallaron, "The ASSIGN Parallel Program Generator," May 1991, Technical Report CMU-CS-91-141, School of Computer Science, Carnegie Mellon University, Pittsburg, PA 15213.

[Hat88] Derek J. Hatley and Imtiaz A. Pirbhai, Strategies for Real-time System Specification, Dorset House Publishing, New York, 1988.

[Hay87] Ian Hayes, et.al., Specification Case Studies, Prentice Hall International (UK) Ltd., Cambridge, 1987.

[Hea91] Walter S. Heath, Real-time Software Techniques, Van Nostrand Reinhold, New York, 1991.

[Hil85] W. Daniel Hillis, The Connection Machine, MIT Press Series in Artificial Intelligence, The MIT Press, Cambridge, Massachusetts, 1985. ISBN 0-262-08157-1.

[Hoa85] C.A.R. Hoare, Communicating Sequential Processes, Prentice Hall, 1985.

[Hoa91] C.A.R. Hoare, "The transputer and occam: a personal story," *Concurrency: Practice and Experience*, Vol. 3, No. 4, pp. 249-264, Aug. 1991. The entire issue is dedicated to transputer applications.

[Hyp86] Proceedings of the First Conference on Hypercube Multiprocessors (Hypercube Multiprocessors 1986), Knoxville, Tennessee, August 26-27, 1985. M.T. Heath ed., SIAM, Philadelphia, 1986.

[Hyp87] Proceedings of the Second Conference on Hypercube Multiprocessors (Hypercube Multiprocessors 1987), Knoxville, Tennessee, September 29-October 1, 1986. M.T. Heath ed., SIAM, Philadelphia, 1986.

[IEE76] IEEE Std. 1076, Very High Speed Integrated Circuit Definition Language (VHDL), Institute of Electronic and Electrical Engineers, New York, 1987.

[IEE83] Software Engineering Standards, The Institute of Electronic and Electrical Engineers, Inc. 345 E. 47th Street, New York, NY 10017, 1983.

[IEEE88] IEEE Standard 982.1-1988, "Standard Dictionary of Measures to Produce Reliable Software," The Institute of Electronic and Electrical Engineers, Inc. 345 E. 47th Street, New York, NY 10017.

[INM88] INMOS Limited, "Specification of instruction set / Specification of floating point unit instructions," in *Transputer Instruction Set - A compiler writer's guide*, pp. 127-161, Prentice Hall, Hemel Hempstead, Hertfordshire HP3 4RG, UK, 1988.

[Jai91] Raj Jain, The Art of Computer Systems Performance Analysis: Techniques for Experimental Design, Measurement, Simulation, and Modeling, John Wiley & Sons, New York, ISBN 0471-50336-3, April 1991.

[Jam87] Leah H. Jamieson, Dennis B. Gannon, and Robert J. Douglass, The Characteristics of Parallel Algorithms, The MIT Press, Cambridge MA, 1987.

[Jef85] D. Jefferson, "Virtual Time," *ACM Transactions on Programming Languages*, Vol. 7, No. 3, July 1985.

[Jef87] D. Jefferson, B. Beckman, *etal.*, "Distributed simulation and the Time Warp Operating System," *ACM Operating Systems Review*, Vol. 21, No. 5, Nov. 1987.

[Jef90] D. Jefferson and P.L. Reiher, "Dynamic Load Management in the Time Warp Operating System," Transactions of The Society for Computer Simulation," Vol. 7, No. 2, pp. 91-120, June 1990.

[Joh89] S. Lennart Johnson and Chieng-Tien Ho, "Optimum Broadcasting and Personalized Communication in Hypercubes," IEEE Transactions on Computers, Vol. 38, No. 9, Sept. 1989.

[Jun90] K.K. Jung, H.T. Nguyen, R. Raghavan, and H.S. Truong, "Massively parallel processors in real-time applications," *SPIE*, Vol. 1246, Parallel Architectures for Image Processing, pp. 107-119, 1990.

[JMa90] J. Markoff "A Superhuman Collapse," *The New York Times*, Jan. 17, 1990, pg. A24.

[Kah91] David K. Kahaner, "Parallel processor for many-body calculations: GRAPE," USENET newsgroup comp.research.japan, posted Oct. 22, 1991. The file "grape.91" is available via anonymous ftp on host cs.arizona.edu. Note: Dr. David Kahaner is a numerical analyst visiting Japan for two-years under the auspices of the Office of Naval Research-Asia (ONR/Asia). The material contained in the aforementioned file is the opinion of David Kahaner.

[Kal91] Karl Kalbfleisch, et.al. "Overview of Real-Time Kernels at the Superconducting Super Collider Laboratory," IEEE Particle Accelerator Conference, San Francisco, CA, 1991.

[Kar88] Anna R. Karlin and Eli Upfal, "Parallel Hashing: An efficient implementation of shared memory," *Journal of the ACM*, Vol. 35, No. 4, Oct. 1988, pp. 876-892.

[Kem90] Richard A. Kemmerer, "Integrating Formal Methods into the Development Process," *IEEE Software*, pp. 37-50, September, 1990.

[Kir83] S. Kirkpatrick, C.D. Gelatt, and M.P. Vecchi, "Optimization by simulated annealing," *Science*, 220, pp. 671, May 1983.

[Kun90] H.T. Kung, "iWarp multicomputer with an embedded switching network," *Microprocessors and Microsystems*, Vol. 14, No. 1, January/February 1990, pp. 59-60.

[Lam78] Leslie Lamport, "Time, Clocks, and the Ordering of Events in a Distributed System," *CACM*, Vol. 21, No. 7, pp. 558-565, July 1978.

[Lam91] Catherine A. Lamanna and Wade H. Shaw, "A performance study of the hypercube parallel processor architecture," *Simulation*, March 1991, pp. 185-196.

[Lee91] Leonard Lee, The Day the Phones Stopped: The Computer Crisis--The What and Why of It, and How We Can Solve It, Donald I. Fine, 1991.

[Lee92] Leonard Lee, "Computers Out of Control," *Byte*, Feb. 1992.

[Lel90] Wm. Leler, "System-Level Parallel Programming Based on Linda (A Call to Standards)," pp. 175-190, Proceedings of Third Conference of the North American Transputer Users Group (NATUG-3), April 26-27, 1990, Sunnyvale CA. A.S. Wagner, Ed. Available from IOS Press.

[Lel90a] W. Leler, "Linda Meets Unix," *IEEE Computer*, Vol. 23, No. 2, February 1990, pp. 43-54.

[Les78] M.E. Lesk and E. Schmidt, "Lex: A Lexical Analyzer Generator," Unix Programmers Manual, Bell Laboratories 1978, Seventh Edition.

[Lev81] N.G. Leveson, "Software Safety: A Definition and Some Preliminary Thoughts," Technical Report #174, Department of Information and Computer Science, University of California at Irvine, Irvine CA 92717 USA, 1981. This report is undated, although the author believes it was written in 1981.

[Lev83] N.G. Leveson and P.R. Harvey, "Analyzing Software Safety," *IEEE Transactions on Software Engineering*, Vol. 9, No. 3, September, 1983, pps. 569-579.

[Lev87] Nancy G. Leveson and Janice L. Stolzy, "Safety Analysis Using Petri Nets," *IEEE Transactions on Software Engineering*, Vol. 13, No. 3, pp. 386-97, March 1987. And "Analysing Safety and Fault Tolerance Using Time Petri Nets," Technical Report #220, 1984, Department of Information and Computer Science, University of California at Irvine, Irvine, CA 92717 USA.

[Lev86] N.G. Leveson and J.L. Stolzy, "Safety Analysis Using Petri Nets," *IEEE Transactions on Software Engineering*, Vol. 13, No. 3, March 1987, pps. 386-397.

[LHA91] J.H. Lala, R.E. Harper, L.S. Alger, "A Design Approach for Ultrareliable Real-time Systems," *IEEE Computer*, Vol. 24, No. 5., May 1991.

[Lip90] Richard J. Lipton and David W. Mizell, "Time Warp vs. Chandy-Misra: A Worst-Case Comparison," Proceedings of the SCS Multiconference on Distributed Simulation, 17-19 January 1990, San Diego CA. *Simulation Series*, Volume 22, No. 2, Jan. 1990.

[Liv85] M. Livny, S. Khoshafian, H. Boral, "Multi-Disk Management Algorithms," MCC Technical Report DB-146-85, Microelectronics and Computer Technology Corporation, 8430 Research Blvd., Austin TX USA 78759, 1985.

[Mac88] Allan Mackintosh, "Dr. Atanasoff's Computer," *Scientific American*, August 1988, pp. 90-96."

[Mar90] J. Markoff, "I.B.M. Workstation Wins Test," *The New York Times*, Section D1, March 16, 1990.

[Mar91] J. Markoff, "American Express to Buy 2 Top Supercomputers," *The New York Times*, Section C7, October 30, 1991.

[Mar92a] J. Markoff, "David Gelernter's Romance With Linda," *The New York Times*, Section 3, pg. 1, January 19, 1992.

[Mar92b] J. Markoff, "Pools of Memory, Waves of Dispute," *The New York Times*, pg. C1, Jan 29, 1992.

[McC89] C. McClure, "The CASE Experience," *Byte*, April, 1989.

[Meh84] Kurt Mehlhorn and Uzi Vishkin, "Randomized and Deterministic Simulations of PRAMs by Parallel Machines with Restricted Granularity of Parallel Memories," *Acta Informatica*, Vol. 21, pp. 339-374, 1984.

[Met53] N. Metropolis, et.al. *J. Chem. Physics*, Vol. 21, 1953, pp. 1087.

[Mil78] G. Mills and J. Walter, Technical Writing, Holt, Reinhart, and Winston, 4th Ed., New York, 1978.

[Mol91] Steven Edward Molnar, "Image-Composition Architectures for Real-time Image Generation," TR91-046, Oct. 1991, The University of North Carlonia, Chapel Hill, Department of Computer Science, CB #3175, Sitterson Hall, Chapel Hill NC 25799-3175.

[Mor84] Carroll Morgan and Bernard Sufrin, "Specification of the UNIX Filing System," *IEEE Transactions on Software Engineering*, Vol. SE-10, No. 2, pp. 128-142.

[Mor85] R. Morrison, S. Otto, "The Scattered Decomposition of Finite Elements," Caltech Concurrent Computation Program Note C^3P 286, May 1985.

[Mus87] J.D. Musa, A. Iannino, K. Okumoto, Software Reliability: Measurement, Prediction, and Application, McGraw-Hill, New York, 1987.

[Mus89] J.D. Musa, "Tools for measuring software reliability," *IEEE Spectrum*, Vol. 26, No.2, Feb. 1989, pp. 39-42.

[Mye92] Ware Myers, "High-Performance computing is 'window into future,' says President's science advisor," *IEEE Computer*, pp. 87-90, Jan. 1992.

[Nar90] K.T. Narayana and Sanjeev Dharap, "Invariant Properties in a Dialog System," *ACM International Workshop on Formal Methods*, 1990.

[Nat89a] Lasse Natvig, "The CREW PRAM Model -- Simulation and Programming," Technical Report no. 38/89, Division of Computer Systems and Telematics, The Norwegian Institute of Technology, The University of Trondheim, Norway, Dec. 1989.Postal Address: UNIT/NTH-IDT, O.S. Bragstads plass 2E, N-7034 Trondhiem, NORWAY.

[Nat89b] Lasse Natvig, "CREW PRAM Simulator -- User's Guide," Technical Report no. 39/89, Division of Computer Systems and Telematics, The Norwegian Institute of Technology, The University of Trondheim, Norway, Dec. 1989. Postal Address: UNIT/NTH-IDT, O.S. Bragstads plass 2E, N-7034 Trondhiem, NORWAY.

[Nel83] Robert A. Nelson, Lois M. Haibt, and Peter B. Sheridan, "Casting Petri Nets into Programs," *IEEE Transactions on Software Engineering*, Vol. SE-9, No. 5, Sept. 1983, pp. 590-602.

[Nic90] J.E. Nicholls (Ed.), Z User Workshop, Proceedings of the Fourth Annual Z User Meeting (Oxford, England), Springer-Verlag, London, 1990. ISBN 0-387-19627-7.

[Nik89] Arvind Rishiyur S. Nikhil, "A Dataflow Approach to General-purpose Parallel Computing," MIT Laboratory for Computer Science, *Computation Structures Group Memo 302*, July 7, 1989, 545 Technology Square, Cambridge MA 02139.

[Oga90] M. Ogata, M.J. Flynn, "A queuing analysis for disk array systems," Computer Systems Laboratory, Technical Report CSL-TR-90-443, Stanford University, Stanford CA 94305-4055 USA, August 1990.

[Pag88] Heinz R. Pagels, The Dreams of Reason, The Computer and the Rise of the Sciences of Complexity, Simon and Schuster, New York, 1988. See Chapter 7, and especially pp. 145-148.

[Pap91] G.M. Papadopoulous, K.R. Traub, "Multithreading: A Revisionist View of Dataflow Architecture," MIT Laboratory for Computer Science, *Computation Structures Group Memo 330*, March 1991, 545 Technology Square, Cambridge MA 02139.

[Par91] C. Park and A.C. Shaw, "A Source-Level Tool for Predicting Deterministic Execution Times of Programs," Technical Report #80-09-12, Department of Computer Science and Engineering, University of Washington, Seattle WA 98195 USA. Also appearing in *IEEE Computer*, Vol. 23, No. 6, June 1991.

[Per91] T.S. Perry, "Special Report: Air Traffic Control; Improving the world's largest, most advanced system," *IEEE Spectrum*, Vol. 28, No. 2, pp. 22-36, Feb. 1991.

[Peru89] Max F. Perutz, Is Science Necessary: Essays on Science and Scientists, E.P. Dutton, New York, 1989.

[Pet81] James L. Peterson, Petri Net Theory and the Modeling of Systems, Prentice-Hall, Englewood Cliffs NJ, 1981.

[Pfl91] M.J. Pfleugl and D.M. Blough, "Evaluation of a New Algorithm for Fault-Tolerant Clock Synchronization," Proceedings of the 1991 Pacific Rim International Symposium on Fault-Tolerant Systems, Japan, 1991. Also, "A New Model and Simulation Tool for Fault-Tolerant Clock Synchronization," *Proceedings of the 22nd Annual Pittsburg Conference on Modeling and Simulation*, Vol. 22, Pittsburgh, May 1991.

[Pot89] A. Pothen, H.D. Simon, and K.P. Liu, "Partitioning Sparse Matrices with Eigenvectors of Graphs," Report RNR-89-009, NASA Ames Research Center, July 1989.

[Pot91] Ben Potter, Jane Sinclair, and David Till, An Introduction to Formal Specification and Z, Prentice-Hall International (UK) Ltd., Cambridge, 1991.

[Pou90] Dick Pountain, "Virtual Channels: The Next Generation of Transputers," *Byte* (European & World edition), April, 1990.

[Pou91] Dick Pountain, "The Transputer Strikes Back," *Byte*, August, 1991.

[Pre86] W.H. Press, B.P. Flannery, S.A. Teukolsky, W.T. Vetterling, Numerical Recipes, The Art of Scientific Computing, pp. 326-27, Cambridge University Press, New York, 1986.

[Qui90a] M.J. Quinn and Phillip J. Hatcher, "Compiling SIMD Programs for MIMD Architecture," pp. 291-296, *Proceedings 1990 International Conference on Computer Languages*, New Orleans LA, March 12-15, 1990.

[Qui90b] M.J. Quinn and Philip J. Hatcher, "Data-Parallel Programming on Multicomputers," *IEEE Software*, Sept. 1990, pp. 69-76.

[Qui91] M.J. Quinn, P.J. Hatcher, A.J. Lapadula, B.K. Seevers, R.J. Anderson, and R.R. Jones, "Data-Parallel Programming on MIMD Computers," *IEEE Transactions on Parallel and Distributed Systems*, Vol. 2, No. 3, July 1991, pp. 377-383.

[Rai89] A.R.C. Raine, D. Fincham, W. Smith, "Systolic Loop Methods for Molecular Dynamics Simulation Using Multiple Transputers," *Computer Physics Communications*, Vol. 55, 1989, pp. 13-30.

[Ram90] P. Ramanathan, K.G. Shin, R.W. Butler, "Fault-Tolerant Clock Synchronization in Distributed Systems, *IEEE Computer*, Oct. 1990, pp. 33-42.

[Raz86] R. R. Razouk, T. Stewart, and M. Wilson, "Measuring Operating System Performance on Modern Micro-Processors," *Performance 86*, pp. 193-202, Association for Computing Machinery, New York, 1986.

[Rob89] B. Robinson, "Grand Challenges to Supercomputing," *Electronic Engineering Times*, Sept. 18, 1989.

[Roe91] William H. Roetzheim, Developing Software to Government Standards, Prentice-Hall, Englewood Cliffs, NJ, 1991.

[Sav85] S. Savitzky, Real-time Microprocessor Systems, Van Nostrand Reinhold Company, Inc., New York, 1985.

[Sch88] Karsten Schwan, "Developing high-performance parallel software for realtime applications," *Information and Software Technology*, Vol. 30, May 1988, pp. 218-227.

[Sei85] Charles L. Seitz, "The Cosmic Cube," *CACM*, Vol. 28, No. 1, pp. 22-33, Jan. 1985.

[Sei90] Charles L. Seitz, "Concurrent Architectures," appearing in VLSI and Parallel Computation, Robert Suaya and Graham Birtwistle, Morgan Kaufamann Publishers, Inc., San Mateo CA, 1990. ISBN 0-934613-99-0.

[Seiz88] J. Seizovic, "The Reactive Kernel," Technical Report 88-10, Department of Computer Science, California Institute of Technology, 1988.

[Sim90] C. Sims, "Computer Failure Disrupts A.T.&T Long Distance," *The New York Times*, Jan. 16., 1990, pg. A1. In the same issue on pg. A24, by J. Barron "Struck Dumb by the Incredible: 1-800-NOTHING," and by M.L. Wald "Experts Diagnose Telephone 'Crash'." C. Sims, "A.T.&T. Pinpoints Cause of Phone Disruption," *The New York Times*, Jan. 17, 1990, pg. A1.

[Ski90] David B. Skillicorn, "Architecture-Independent Parallel Computation," *IEEE Computer*, Dec. 1990, pp. 38-50.

[Spi88] J.M. Spivey, Understanding Z: A Specification Language and its Formal Semantics, Cambridge University Press, 1988.

[Spi90] J. Michael Spivey, "Specifying a Real-time Kernel," *IEEE Software*, pp. 21-28 September, 1990.

[SSD91] Paul A. Swatman, Paula M.C. Swatman, Roger Duke, "Electronic Data Interchange: A High-level Formal Specification in Object-Z," *Proceedings of the 6th Austrailian Software Engineering Conference*, Sydney, July, 1991.

[Sta88] J.A. Stankovic and K. Ramaritham, Hard Real-time Systems, IEEE Computer Society Press, Washington D.C., 1988. Also, J.A. Stankovic "A Serious Problem for Next-Generation Systems," *IEEE Computer*, pp. 10-19, Oct. 1988. Available as COINS technical report 88-06 from The Department of Computer and Information Science, University of Massachusetts, Amerherst, MA 01003. Jan. 6, 1988.

[Sta90] J.A. Stankovic and Krithi Ramamritham, "What is Predictability for Real-time Systems?", *Real-time Systems*, Vol. 2, pp. 247-254, 1990.

[Ste88] R.M. Stein, "T800 and Counting," *Byte*, Nov. 1988.

[Ste90] Richard M. Stein, "A Multicomputer Poisson Equation Solver," *Proceedings of the Fourth Annual Parallel Processing Symposium*, Vol. 1, pp. 541-553, Fullerton California, April 4-6, 1990. Proceedings available from the IEEE Computer Society Press.

[Ste91] R.M. Stein, "Real Artificial Life," *Byte*, Jan. 1991.

[Ste92a] Richard Marlon Stein, "The Open Channel -- How to Forestall Electronic Mutually Assured Destruction," *IEEE Computer*, March, 1992.

[Ste92b] R.M. Stein, "Terabyte Memories with the Speed of Light," *Byte*, Mar. 1992.

[Ste92c] Richard Marlon Stein and Philip Presser, "Scalable Concurrent Visualization System," U.S. Patent and Trademark Office, Application # 07/896057, June 9, 1992.

[Sti90] G. Stix, "Cold Cathodes," *Scientific American*, Oct. 1990, pp. 122-23.

[Str86] Bjarne Stroustrup, The C++ Programming Language, pg. 127, Addison-Wesely Publishing Company, Inc., 1986.

[Sub91] Jaspal Sublock and H.T. Kung, "A New Approach for Automatic Parallelization of Blocked Linear Algebra Computation," *Proceedings of Supercomputing '91*, Albuquerque NM, Nov. 1991.

[Swa91] Paul A. Swatman and Paula M.C. Swatman, "Is the Information Systems Community Wrong to Ignore Formal Specifications Methods?," *Proceedings of the Conference Shaping Organizations, Shaping Technology*, SOST 1991, Adelaide, Oct. 1991.

[Tan92] Andrew Tannenbaum, Modern Operating Systems, Prentice-Hall, Englewood Cliffs, NJ, 1992.

[Til88] R.F. Tilton Jr., U.C. Singh, I.D. Kuntz Jr., and P.A. Kollman, "Protein-Ligand Dynamics: A 96 picosecond Simulation of a Myoglobin-Xenon Complex," *J. Mol. Biol.*, Vol. 199, pps. 195-211, 1988.

[Tra91] K.R. Traub, G.M. Papdopoulos, M.J. Beckerle, J.E. Hicks, and J. Young, "Overview of the Monsoon Project," MIT Laboratory for Computer Science, *Computation Structures Group Memo 338*, July 1991, 545 Technology Square, Cambridge MA 02139.

[Tro90] Trollius User's Reference and Trollius Reference Manual for C Programmers, by the Advanced Computing Research Institute of the Cornell Theory Center 265 Olin Hall, Ithaca New York, 14853-5201, and the The Ohio State University, Research Computing Center, 1224 Kinnear Road, Columbus OH, 43212.

[Tur36] A.M. Turing, "On Computable Numbers, with an Application to the Entscheidungsproblem," *Proceedings of the London Mathematical Society Series 2*, Vol. 42, pp. 230-265, 1936; correction, *ibidem*, Vol. 43, pp. 544-546, 1937.

[Ung88] Brian Unger and David Jefferson, Distributed Simulation 1988, Proceedings of the SCS Multiconference on Distributed Simulation, Feb. 3-5, 1988, San Diego CA. *Simulation Series*, Vol. 19, No. 3, July 1988.

[Val90a] L.G. Valiant, "General Purpose Parallel Architectures," Chapter 18 of Handbook of Theoretical Computer Science, J. van Leeuwen Ed. Elsevier Science Publishers B.V., 1990.

[Val90b] Leslie G. Valiant, "A Bridging Model for Parallel Computation," *CACM*, Vol. 33, No. August 1990, pp. 103-111.

[Val90c] Leslie G. Valiant, "Bulk Synchronous Parallel Computers," Harvard University Center for Research in Computing Technology, TR-08-89, April 13, 1989, Aiken Computation Laboratory, 33 Oxford Street, Cambridge MA 02138.

[Wat91] G. F. Watson & R. Thomas, "TCAS Sees Ghosts," *IEEE Spectrum*, August 1991, pg. 58. Another recently related report is by John R. Cushman, "F.A.A. Seeking Replacement of Faulty Plane Signal Device," *The New York Times*, April 25, 1992, pg. Y9.

[Web91] Jon A. Webb, "Adapt: Global Image Processing with the Split and Merge Model," Technical Report CMU-CS-91-129, July 1991, School of Computer Science, Carnegie Mellon University, Pittsburg, PA 15213.

[Wel87] P.H. Welch, "Managing Hard Real-time Demands on Transputers," *Proceedings of the 7th Occam Users Group*, Grenoble France, Sept. 1987.

[Wex89] John Wexler, Concurrent Programming in Occam 2, Ellis Horwood Series in Computers and Their Applications, Ellis Horwood Ltd., West Sussex, England, 1989. ISBN 0-7458-0394-6.

[Wil90] Roy D. Williams, "Performance of Dynamics Load Balancing Algorithms for Unstructured Mesh Calculations," California Institute of Technology Concurrent Computation Project Report C^3P 913, June 1990. A similar work will appear in a forthcoming volume of *Concurrency: Practice and Experience*.

[Win90] J.M. Wing, "A Specifier's Introduction to Formal Methods," *IEEE Computer*, Vol. 23, No. 9, September 1990.

[Wit91] E.E. Witte, R.D. Chamberlain, and M.A. Franklin, "Parallel Simulated Annealing Using Speculative Computation," *IEEE Transactions on Parallel and Distributed Systems*, Vol. 2, No. 4, Oct. 1991, pp. 483-494.

[Wol83] Jeremy M. Wolfe, "Hidden Visual Processes," *Scientific American*, Feb. 1983.

[You89] Edward Yourdon, "The Yourdon Approach," *Byte*, April, 1989, pp. 227-230.

Date of Birth: March 25, 1961.
Birthplace: Los Angeles, California USA.
Education: Corona del Mar High School (Newport Beach CA); B.S. Physics, University of California at Irvine awarded in 1984.
Hobbies: Downhill and cross-country skiing, cardiovascular exercise, reading, creative writing.
Awards: UC Irvine Alumni Association Distinguished Student Scholar in Physics 1983 and 1984. NSF 91-20 Phase I SBIR for "Improved Visualization System for Massively Parallel Processors."
Currently Resides: Santa Clara, CA.

Trademark Acknowledgement

Intel and iWARP are trademarks of the Intel Corporation.
UNIX is a registered trademark of UNIX Systems Laboratories.
Occam is a trademark of the Inmos Group of Companies.
i860 is a trademark of Intel Corporation.
CM-5 is a trademark of Thinking Machines, Inc.
KSR1 is a trademark of Kendall Square Research, Inc.
Linda is a trademark of Scientific Computing Associates, Inc.
Trollius is a trademark of The Ohio State University and Cornell University.

Index

L.-Brault